STRIKING BACK AT
STROKE

STRIKING BACK AT

STROKE

A Doctor-Patient Journal

Cleo Hutton and
Louis R. Caplan, M.D.

THE
DANA
PRESS

New York | Washington, D.C.

Library of Congress Cataloging-in-Publication Data
Hutton, Cleo, and Caplan, Louis R.
Striking back at stroke : a doctor-patient journal / by Cleo Hutton, Louis R. Caplan.
    p. cm.
Includes bibliographical references and index.
  ISBN 0-9723830-1-8
  1. Caplan, Louis R.--Health. 2. Cerebrovascular disease--Patients--United States--Diaries.
3. Cerebrovascular disease--Popular works. 4. Nurses--United States--Diaries. I. Hutton,
Cleo, 1949- II. Title.
    RC388.5 .C3298 2003
    362.1'9681--dc21

                                      2002155047

ISBN: 0-9723830-1-8

THE DANA FOUNDATION
745 Fifth Avenue, Suite 900
New York, NY 10151

THE DANA PRESS
The Dana Center
900 15th Street NW
Washington, D.C. 20005

www.dana.org

DEDICATION

THIS BOOK is dedicated to everyone who has had a stroke, their families, and loved ones who have learned or are in the process of learning that our spirit endures. Also, with sincere gratitude to physicians, neurologists, cardiologists, cardiac surgeons, nurses, aides, clergy, stroke rehabilitation therapists, college access services, college professors, rehabilitation counselors, friends, and family. Without their help this text would not have been written.

—*Cleo Hutton*

I DEDICATE THIS BOOK to all of my stroke patients and their loved ones. They have taught me so much about stroke, medicine, courage, ingenuity, perseverance, and caring.

—*Louis R. Caplan, M.D.*

IN MEMORY OF

IN MEMORY OF all the people who have died from stroke. We hope this book will provide greater understanding of brain attack, so that someday, in the near future, there will be a cure. Also in memory of Cleo's brother and her friend Bill.

Contents

The authors express sincere appreciation and thanks to The Dana Press, publisher for The Dana Foundation, whose principal interest is brain research, and Jane Nevins, editor in chief, for her guidance and encouragement.

Cleo Hutton wishes to thank Louis R. Caplan, M.D. for his insight and the time he dedicated to writing and editing this book, as well as reviewing the illustrations. Our unique collaborative effort is aimed at assisting stroke survivors, their families, and the medical community.

Cleo Hutton also acknowledges with gratitude the assistance of her friend Alice Percy, who helped edit the early drafts of the text in 1994.

Foreword

CLEO HUTTON

SOME PEOPLE never get the opportunity to see the inner workings of the brain. For many of us, the anatomical and biological workings of our bodies just happen—like the ticking of a clock. The clock quietly ticks and we go about our day. But what happens when the alarm rings and we lose the ability to control our bodies and our thought processes? We understand the essence of our existence—personalities, memories, knowledge, hopes, and dreams—is contained within our brains. After a stroke, we are jolted into realizing that many other functions of our bodies are contained between our ears as well.

This journal is a day-to-day record before, during, and after a powerful, life-altering event—stroke. Due to the severity of the stroke, I lost my ability to communicate. Slowly, I began to write a word, a thought, and eventually a complete sentence. The process of deliberately picturing each letter while trying to remember how to write, then continuing to construct a simple sentence became an agonizing chore. Language known from childhood and vocabulary stockpiled in my brain seemed to dissolve before I could send speech from my mouth. It took months to regain the ability to speak clearly and logically. Even today, ten years later, there are times my language ability is gauged by the energy level of my brain. In time, after relearning to formulate a few words, I began dictating my thoughts into a tape recorder. Listening to the tape recordings over and over again, I heard the constant repetition and fragmented thought patterns I had expressed. For several years post-stroke, my small tape recorder served as a constant companion, as I openly and with the consent of those who were speaking to me, taped vital conversations my injured brain did not have the capacity to recall.

For most of my life, I have kept a journal, but it had never been as imperative as after the stroke. As a nurse, I felt compelled to document the medical aspect of my recovery. As an adult who had lost control of communicating and understanding the written and spoken word, I needed to express the terror and frustration associated with stroke. While in the hospital after the stroke, I began to painstakingly write this journal. The original journal entries are handwritten in a blue spiral notebook with pictures of my three children taped to the cover.

The purpose of publishing the journal is to help other stroke survivors and their families. I also hope to aid in stroke awareness—to inform the public of the warning signs of stroke (cerebral vascular accident) by describing events that led up to this medical emergency. I hope to indicate some life lessons I have learned—to help fellow stroke survivors in their journey toward recovery by exposing a particularly vulnerable period in my life. Originally, the journal was my tool for memory. Later, it served to reassure and encourage each slow step in the healing process. The journal contains instances of failure and backsliding, frustration and tears, but my desire is that you, the reader, will also see the triumphs and spirit to survive and grow through it all.

The journal entries in *Striking Back at Stroke* are the original text with very little change. The most difficult task I faced in preparing this book was resisting the temptation to improve the writing. Language problems were a reality in my stroke and recovery that I wanted to indicate to the reader, so the difficulty I experienced in word-finding and placement skills shows, but the recovery of some of these abilities over time is evident also. In the hope of bringing the reader as close as possible to my experience, I changed the verb structure to present tense. That was challenging because stroke deficits continue to plague my writing skills, especially in conjugating verbs and the use of present- or past-tense structure. And, because I know the characters who played a significant part in my life, I have added detail where my hand did not add enough record for a public format. To respect the privacy of family members and friends, their names have been changed. Because I am only one person among many hundreds of thousands of stroke survivors, I am also writing under another name.

The process of bringing the book to publication taught me, too: it was a therapeutic way for me to remember the tragedies, as well as the joys, of life.

ON JUNE 9, 1992, at the age of 43, I had the stroke, followed nine days later by another. The type of stroke I had is called "ischemic," referring to inadequate blood flow to some parts of the brain. Because the strokes were caused by a congenital heart defect, within a month I had heart surgery. Although this is my personal account of emotional and physical recovery, more than 750,000 people are affected each year by stroke in the United States. It is the third leading cause of death and the leading cause of major disability in this country. Stroke leaves challenging deficits; simple tasks become, at times, extremely difficult to perform. Over the years, I have grown by learning to adapt to a new way of living.

After the stroke, questions constantly plagued me. As a nurse, why didn't I see the warning signs? As a middle-aged person, how could someone so young have a stroke? With no family history of stroke, why me? What part of my brain was affected? What is the meaning of the word *recovery* as it relates to stroke? Could it happen again?

During rehabilitation, I attempted to answer some of these questions by seeking information in medical textbooks. I found a book—*Stroke: A Clinical Approach*—written by Louis R. Caplan, M.D., of Boston, Massachusetts. This book was written for the medical community, but because of stroke deficits,

I could not understand even the basic medical terminology I had once known. I knew, however, that he held key information to understanding strokes. In 1994, I sent him a letter—full of syntax errors and misspellings that I was not aware of making—requesting his help in finding answers to my many questions.

Dr. Caplan responded, and we decided to collaborate on this book. From there, it took several more years of stroke healing and education before I could even imagine finding a publishing home for our manuscript. In 2002, ten years after the event—yet with ten years of stroke recovery, experience, and a wealth of medical knowledge from Dr. Caplan—we give you an inside look at stroke.

THE EFFECTS OF STROKE are challenging. My challenge to the reader is to gain knowledge so that together we may prevent stroke. And while we await significant research developments toward better treatments, I challenge the reader to face stroke head-on, to assist persons who have experienced a stroke to reach their full potential, to assist families who are providing care and support, and to help improve outreach measures so that every stroke survivor may have the best chance for recovery.

LOUIS R. CAPLAN, M.D.

AT THE BEGINNING of this collaborative work, I want to introduce myself and tell you something about how I came to work with Cleo on her book. I am a doctor who has devoted his entire medical career to stroke. After my medical and neurology training in the 1960s and two years in the army as a physician, I spent a year as a fellow in stroke and cerebrovascular disease at the Massachusetts General Hospital in Boston and then embarked on my first hospital staff experience in 1970 at the Beth Israel Hospital in Boston. There I was one of the founders of the Harvard Stroke Registry. This was the first time doctors collected detailed information about each stroke patient and created a computer database.

After eight years at Beth Israel, I accepted an offer to become chairman of the Neurology Department at Chicago's prestigious Michael Reese Hospital and professor of neurology at the University of Chicago. But I missed New England, and in 1984, taking my wife and six children, I went back to Boston, as neurologist in chief of the New England Medical Center and professor and chairman of neurology at Tufts University. Fourteen years later, I closed the circle, returning to where I'd started, Beth Israel (now the Beth Israel Deaconess Medical Center), and my current position as chief of the stroke service, and professor of neurology at Harvard Medical School. These two institutions, Beth Israel Deaconess and Harvard, are among the leaders of clinical stroke research in America.

At each of the hospitals where I worked, I headed the stroke service, caring for and consulting on all stroke patients in the hospital. At each hospital, we made a computerized stroke registry to record and keep track of all stroke patients and to learn about the patients, their diseases, and their findings and problems. In this way we hope to improve our treatment of future stroke patients. I have also held leadership positions in the American Heart Association, the American Neurological Association, and the American Academy of Neurology. I was presented with the Distinguished Achievement Award of the American Heart Association for my work for stroke patients and was given a Distinguished Faculty Award by Tufts University School of Medicine and the Golden Key Alumni Award by my medical school.

Cleo found my name by looking up "stroke experts." She had also read my book *Stroke: A Clinical Approach*. She asked me to review and edit what she had written about her own stroke. She hoped to publish her story so that people could learn from it. I was fascinated by her experiences. More important, I saw her story as a way to help people understand more about stroke and its causes and the way doctors diagnose and treat it. I asked her if I could help write the story, interweaving more general ideas and concepts with her account. She agreed and we have been partners in this work since.

Although I have written or edited twenty-six technical books for doctors, this is only the second book I have contributed to that was written for patients, their families and loved ones, and the general public. The other is the American Heart Association's *Family Guide to Stroke Treatment, Recovery, and Prevention.* I hope to make all of my remarks plain and practical and yet detailed enough to be helpful. The book, however, is about Cleo's stroke and her harrowing experiences. I was brought in afterward, an invited guest asked to interpret various aspects of her strokes and her story. I know that everyone can learn from her experiences, and I want to place her story in a broader context of stroke and of medicine and illness in general. In particular, I want readers to gain knowledge about the brain and how it looks and works so that they can understand the symptoms of stroke and its related disabilities. I also want them to have some understanding of what a stroke is, what causes strokes, how strokes are diagnosed, and how strokes are treated. Although Cleo's stroke occurred ten years ago, it came after a period of real improvement in the treatment of stroke, and we use the same practices and technology now. We continue working to refine these tools and treatments, however, and I will mention recent advances in this book. In my Afterword, I will tell you about new approaches on the horizon.

Stroke is a very common condition. About three quarters of a million Americans have a stroke each year. At any one time there are about 2 million people living in the United States who have had strokes sometime in their past. Stroke is the third leading cause of death in the United States and many other countries in Europe and Asia. Strokes develop in more women each year than breast cancer. Cleo was 43 years old at the time of her stroke. One widely held but mistaken idea is that strokes affect only very old people. Young adults, adolescents, children, and even babies can have strokes. Because of advances in our understanding of stroke and its causes, the frequency of stroke and its severity seem to be declining. According to the American Heart Association, the death rate for stroke dropped from 89 per 100,000 people in 1950 to 30 per 100,000 in 1988. But even though fewer strokes kill outright, their aftermath is a greater challenge now. A 2002 study of long-term disability after a first stroke revealed that four in ten survivors live five years, but one third of them remain disabled and one in seven requires permanent institutional care at the five-year mark.

Not only do strokes cause symptoms and disabilities in the individual patient, but that person's problems have a profound impact on his or her family, friends, coworkers, and community. In patients with serious medical disorders, the illness and the patient interact in so many ways that the two are often not separable. The illness is not understood without knowing the patient well. In this book, we will learn much about Cleo and the people and events in her life. We will also learn how her various medical problems became interwoven with her life story and that of her loved ones.

CHAPTER 1

Brain Attack

I BEGIN WHERE I knew something very strange was happening. Before the journal entries, I remember events that signaled trouble. This recollection came only from healed long-term memory:

In the spring of 1987 our oldest daughter, Paige, was in a school play. She was Daisy Mae in an adaptation of *Li'l Abner*. I drove to a nearby flower shop to get a bouquet of daisies for her opening night. But when I got to the store, less than two miles from our home, I couldn't remember the original intent of my errand. All the flowers melded together, and I became disoriented and began to order large quantities of flowers. Quickly, I realized that it must have been the excitement of the moment and recalled my original plan to purchase a simple bouquet of daisies. Upon leaving the store, I got in the car and began to drive home. Somehow—I don't remember exactly how because I was concentrating on the act of driving and keeping the car on the road—the bigger picture of where I was going vanished from my memory. I drove into the parking lot at a shopping center that looked familiar, but I could not recite its name, even with the bright red letters that hung in a half-moon shape above the roof. Stepping on the brake, I put the car into park and tried to get my mind under control. "Now Cleo, you know where you are! Stop this! What in the hell is the matter with you? You take that street down to Central Avenue. Oh, what Central Avenue? C-e-n . . . Wait! You know this! Stop and think!" I muttered to myself. I remember telephoning my general practitioner's office from a pay phone nearby.

"I'm lost! I don't know where I am, but I'm near your office. Oh, this is ridiculous!" I sobbed.

I remember talking to my doctor, but I can't recall hearing his soothing voice. I was embarrassed about making the silly call, and, with a "Please forgive me, I'm fine," I hung up and called my husband, Larry. I knew that he'd be home on Saturday. Trying to reassure myself that it was indeed Saturday and finally making the connection that the daisies in the car had to do with Paige's performance as Daisy Mae, I attempted to relax when the phone rang at home. Quickly peeking at my watch, I wasn't sure if it was twelve-thirty or six o'clock—it couldn't be that late! The play starts at—

"Larry, I'm lost," my voice quivered.

"Where are you?" he inquired.

"I'm at a shopping center and there's a theater and a film shop and a restaurant nearby. You know, the one we take the kids to after church on Sunday—the one with the chocolate chip pancakes."

"Oh! What are you doing at City Plaza?"

"Yes, City Plaza! That's were I am! Will you come and get me?"

"Huh? You've got the car!"

"Oh, of course! Never mind. I'll be home shortly." I hung up the phone and began to weep. I cried about the horrible mistake I had made by sounding so idiotic. I cried as I realized I had been overwhelmed with stress. I didn't think I was stressed on a warm and cheerful Saturday afternoon. I sat in the car and waited for the moment to pass. Eventually it did and I drove home without further incident.

When I got home Larry asked what the problem was, but I just shrugged it off with an "I'm fine, just got turned around" answer. This is one of my first recollections of a TIA (transient ischemic attack).

On another occasion, I couldn't read the menu when we went out to dinner with friends and ordered "chi-king or-a-gan-o" instead of chicken oregano. Everyone laughed at my faux pas and I did too. The words looked familiar, but I couldn't say them. Again, I had had a TIA without realizing the significance of the event.

Another time, Larry and I were out with another couple at a restaurant and were sitting in a booth. The shoe came off my left foot, and I couldn't retrieve it with my foot. The left side of my body became heavy, and I began to lean on Larry, who was sitting beside me. We laughed and thought it was the effects of too much alcohol, but I had just ordered my first drink of the evening. It passed, but I remember an overwhelming sense of fatigue, and I had difficulty walking out to the car. This marked my fourth or fifth TIA. I was unconsciously playing a game of Russian roulette with my brain. I never wrote of these incidents in my journal, as I thought they were too ridiculous to mention. Not until several years post-stroke, which seems like a lifetime later, can I recall with clarity these telltale events that occurred before May 1992.

LARRY AND I WERE MARRIED in 1971. We lived in a metropolitan city in our northern state, a couple of hours away from our hometown and families. We had moved there early in our marriage to follow Larry's goal of earning a master's degree. We had left behind an array of relatives, including his parents, sister, and brother-in-law and my two brothers and three sisters and their families. I had taken a job at a nearby university hospital, but things hadn't gone well for Larry. He had quit graduate school and begun looking for employment. While unemployed, Larry became the epitome of Mr. Mom. He rocked our little daughter, Paige, born in 1973, potty trained her, and kept the house in tip-top condition. A gangly guy—at six feet six, a foot taller than I—all legs and arms, he held tea parties on a tiny table set up with lace doilies as napkins, and crackers and cheese for Princess Paige and Giant Larry. He eventually found work in sales and discovered he had a calling for it.

I am a licensed practical nurse, state board certified in June 1970. Most of my practical experience was in pediatric nursing—I worked in a specialty hospital and I was a specialty nurse. I continued working in that capacity for several years after Larry and I were married. After Paige was born, I missed being at home with her. My heart was torn between the very sick children I cared for and our healthy, happy child at home. When at work, I didn't think about home, but at home all I could think about was the sick children in the hospital. I cared for premature infants, tube-feeding them, watching the rhythms of their tiny hearts, and carefully weighing them, charting every ounce of growth. I cared for extremely ill children, and I rocked babies that were born addicted to drugs their mothers had taken during pregnancy, watching as the infants fought off the addiction. The children in my care became part of me, just as if they were my own. I finally resolved my dilemma by bringing together home and child care: I began a licensed child care business in our home. This solution was so satisfying that in the late 1970s, I returned to school at night and graduated with a child development teaching degree. Our second child, Mark, was born in 1977, and our third, Betty Rae, in 1979. Parenting was a gift I excelled at.

The late 1980s were rough financial years for Larry and me. We sold our home after living there for many years and purchased another, about a mile or so away from the first. The deal went sour—our new house, which had been lived in previously, turned out to have structural defects and was condemned by the city—and we lost our equity. Like nomads, we scoured the area for suitable rental housing for a couple with three children and a dog. It wasn't easy, but eventually we secured a little green house on a corner by the city park and a beach. A few years earlier, Larry and a friend had become partners and started their own business in sales, so this could not have happened at a worse time. Because we didn't have a house of our own, I had to suspend my child care business and get another job. Our marriage had a solid foundation but was suffering from stress fractures. We'd rally and work as a team to quickly mortar up the damage we sustained both financially and emotionally, but all of us missed our old home. We played a silent game of blame, never once outright mentioning that each of us had wanted to stay in our former home. Instead, Larry and I would take long strolls around the neighborhood in the evening and reflect on the past. We'd speak of "the good ol' days" in the former house and laugh about how Larry had built several gardens for me, all in six-by-three-foot treated lumber beds, and how they resembled graves during the fall and winter months. This was our futile attempt at humor, as Larry's employment included sales to funeral homes.

After we finally resolved the real estate fiasco, we managed to obtain financing to begin building a house. By 1988, we were settled in our new home, Larry had taken a new sales position with a national company, and I had resumed providing child care in a part of the house we had had designed specifically for that purpose.

Our children were adolescents and required all the necessities and accoutrements that go with school, extracurricular activities, and being cool. Larry's career was going well, but it required him to travel throughout the state. It was not unusual for him to leave on road trips before the sun rose and be away for a few days each week. Financially, we had recovered enough to keep the bases covered and buy a car for Larry's travels. I began saving for retirement in an IRA; Larry had another IRA in addition to his company's retirement plan.

I was twenty-five pounds overweight shortly after moving into our new house. I purchased a few videotapes, but my energy level was rapidly decreasing, and I'd end up sprawled on the couch watching Jane Fonda stretch and "feel the burn." But after all we had been through, I felt that life was good again. I didn't realize the framework of our marriage could yet be fragile. It would take the stroke to shatter it.

AT THE TIME of my stroke, in 1992, our oldest daughter, Paige, was 19 and busy with her first year of studies at a community college. Additionally, she was working as an intern at the local newspaper office and had a job at a local clothing store. We were so proud of her; she earned her own byline in the newspaper with a humorous article about women and car repairs. Our son, Mark, was 14 and trading in his childhood bookishness for "mullet" haircuts (crew cut on top and long in the back) and rock music. Mark, who appears to be the mirror image of Larry—oval face, broad shoulders, and lanky—is different from his father. Larry is gregarious, and Mark is shy. Larry is tense and can be explosive, while Mark is calm. Mark exuded an inner peace that is very rare with adolescent boys, and he chose his words and his friends very carefully. Betty Rae was 12, the youngest but feistiest of our family. Behind closed doors she was an independent thinker and attained excellent grades in school, but publicly she insisted on being part of every aspect of her peer group. Betty Rae usually ate breakfast with me and the day care children, read stories to them, and assisted in outdoor activities. She was changing like a chameleon every day; she had a list of needs a mile long, but when her needs were met, she immediately changed her mind. Betty Rae was a typical adolescent, going through growing pains.

April and May 1992   My older brother, Brian, is very ill and his T cell count is extremely low. A few months ago the HIV developed into AIDS. He is so thin he can barely walk, and his face appears concave and gaunt, with only his big blue eyes to preserve his spirit.

There is no time to journal with any frequency now. Every weekend I drive about 160 miles to my brother's home to relieve my sister from her devoted duties of home hospice care. During the last trip back home, my depth perception seemed a little off: I couldn't read the highway signs until I was practically on top of them, and I had a bit of a startle when another driver honked the horn while trying to pass on my left. I probably need my glasses checked.

Mid-May 1992 The weather promises a spring morning, but Larry and I are sitting in blue vinyl chairs in a waiting area of the hospital's emergency room. I have a sharp, constant pain behind my eyes, and I find it difficult to focus my vision. The left side of my body is heavy, and I cannot hold my head upright. I cannot answer the barrage of questions being asked at the admission desk, so I quietly slide my insurance card over to the woman sitting behind the desk.

When my name is called, I begin to feel slightly improved; the left arm weakness is almost gone by the time the doctor enters the examination room.

"And what seems to be the problem?" asks the doctor, glancing at the chart he retrieved from the plastic case on the door.

"It must be those cluster migraines again. I took the prescribed medication when it began, but it seems worse this time. I'm dizzy . . . I don't know." I begin to lean against the examining table with an overwhelming urge to lie down, but the doctor asks me to stand up and place my hands out in front of me. My left hand appears slow and jerky, and my left arm feels as if I hit my elbow and it stings as if it were asleep.

"I think we should admit you for observation. I'd like to have a neurologist take a look at you."

"I feel better now! It was so strange. The pain began suddenly and seemed to dissolve."

"I'd like to run some tests and get an MRI. Have you had any other symptoms?"

What type of symptoms? "A few nights ago I had a tremendous headache, one that forced me to bed with a cold compress on my forehead. The pain was so severe I had trouble moving. I just wanted to sleep it off."

Next day *[This journal entry has no exact date. Only now do I realize this omission is extremely rare in my lifetime journal entries.]*

After a day in the hospital, and a MRI scan of my head, I am sent home. Over the next few days, I take an aspirin a day, as the doctor prescribed. Although the doctor did not explain why he prescribed it, I assume the aspirin is to prevent further headaches.

My child care business is suffering. The daily twelve-hour schedule necessary to meet parents' needs is pushing the limits of my health. An absence not only hurts us financially, but starts a domino effect with my clients, as they have to find a substitute provider. Caring for twelve children demands high energy, and I don't have the strength it takes to lift a pencil. Last night, I called my clients to request today off. I am sure this lack of pep is just stress, so I try to get lots of rest.

My general practitioner says that staying in bed indicates depression. I have never done this before, and I cannot understand why I am so tired. No, it is not depression. I have to lay my head down because my body is weak. When I sleep this heavy, groggy sleep, my mind seems to be in an abyss, and it is difficult to wake. I have never experienced this before.

FOLLOW-UP DOCTOR APPOINTMENTS.

May 1992 While at my checkup, I ask my doctor about the persistent symptoms of tiredness, and the numbness of my left arm and leg. Brushing away past incidents that seem irrelevant to me, I try to think in medical terms and recite my list of symptoms to the physician. He refers me to an internist. The internist refers me to a cardiologist. The cardiologist does an ultrasound and a stress test and says I am fine. I return to the internist, and he feels a node in my neck and refers me to an endocrinologist for thyroid surgery. After I have a thyroidectomy, my health does not improve.

Intermittently, my left hand and arm become numb and awkward, but it goes away within a few minutes. At times, while reading to the day care children, my voice sounds disjointed, as if in slow motion, and I seem to daydream while whimsical, melodious words sound like nonsense. Quickly, I dismiss this thinking and opt for the belief that it is impossible for me to stay enthusiastic about the same children's book throughout every reading.

Sometimes I am unsteady on my feet or feel as if I am falling. I cannot play with the children as I have previously done. Maybe my age hinders my agility.

This weird menagerie of events occurs infrequently and not particularly at the same time. I do not have a fever. My speech is slow and I purposefully avoid family conversations because their staccato and half-sentence speech bother me. I am becoming easily frustrated and I need time to catch on to things. I have no energy to start new projects. Maybe it is all in my head.

THE STROKE. Just before dawn I hear the screech of the garage door as it opens and closes when Larry leaves for work. I remember him saying he had to drive a long way and wanted to get an early start around five-thirty. I might as well get up and not wait for six-thirty.

| June 9, 1992 |

I don't know why I slept on the couch in the family room last night. Maybe I was too tired to climb the stairs to our bedroom. Maybe I just had to be alone. I am feeling so sad, alone, isolated, and miserable these days that I do not want to inflict my inner terror on anyone else.

I reach for my watch, but my arm and hand will not move. I must have slept on my arm. Soon the prickly sensation will begin. I am extremely tired; maybe a few more winks will not hurt. I can see the digital clock. Every fifteen minutes I try to get up but I find the same problem, only much worse with every attempt. It is six-thirty and I have to get up. By using all the energy my body has, I stand up, urinate on myself, and fall.

I drag myself to the telephone, pull it down to the floor, and push the button on speed dial for our general practitioner. His answering service reaches him and he quickly returns the call. He says to call an ambulance and go to the hospital immediately. He will have a neurologist meet me there.

The main unit of our household intercom system is located close to the telephone, and the red light indicates that it is working. I summon Paige from her upstairs bedroom. "Mom, your voice sounds so strange!" she answers. I try to tell her to take me to the emergency room. Time seems to wait, slowly ticking off the seconds like hours as I hear the hair dryer. Is she possibly getting ready for class? But I hear her walking in the upstairs hall.

I don't want an ambulance for stress-related symptoms. My child care clients will be arriving soon. Paige's footsteps are on the stairs now and she continues to come closer to me down the hall and into the kitchen. I can hear her shoes clap against the linoleum floor. "Mom? Mom!" she shrieks, but somehow her voice sounds muffled. I remember smelling something, maybe the scent of her body lotion next to my face. Words go in and out, on and off, pictures, snapshots of her concerned face. My body is heavy and won't move at will. I want to sleep. Paige is slim, approximately 130 pounds, give or take an ounce, and is five feet nine. She disappears around the corner toward the attached garage and opens the back-seat car door first, then the garage door, and returns for me. Again, I am overpowered by the scent of something comforting and her soft skin against mine. Her movements seem frantic, but it may be only in my mind. Her breathing is accelerated. "Hold me, Paige! Don't drop me! I think I can help." I'm asleep, awake, aware, yet unaware. As Paige carries me to the car, the clients are arriving. I hear Paige say, "I've got to take Mom to the hospital. I'm sorry, no child care today."

I fall asleep in the car in my wet nightshirt. Fragments of pictures enter my mind. Then I dissolve into sleep—not a dreaming sleep, just a deep emptiness. A wheelchair, doors open, I cannot hold my body erect, now I am moving, and there are voices that say nothing. The smell of Paige mixed with hamburger wrappers from the car is replaced by open spaces and cleaning fluids that sting my senses. Somehow I feel safe, the hospital personnel know what to do.

Paige stays with me and calls home to tell Mark and Betty Rae that we are at the hospital. Only now that I think back can I comprehend the reason Paige got up so early this particular morning. She had a college final exam scheduled. Larry is driving toward his appointment, somewhere on the highway heading north. He always telephones the family in the evening.

Time means nothing to me. I feel no pain. I am in a warm, white aura of nothingness. I don't worry, I don't cry out. I am safe, secure, and have no conscious knowledge of my family or my own being. I like it and I want to stay. Whatever has happened is over. Without my sense of self, I cannot hear, speak, see, feel, or think.

Later, Paige tells me that there were various medical personnel around me in the little cubicle, yelling at me, pinching, prodding, and shouting at me to wake up. I did not wake up.

A man in a white coat, possibly the neurologist, is saying something, but his voice is gruff as if he is worried. He admits me to the hospital, and later I hear that the CT scan is still showing negative results. He assigns a psychiatrist to my case. I am awake now but not coherent. The next day, the psychiatrist visits with me and asks me several questions. I cannot remember where I am or what is happening. I cannot comprehend what he is saying or why he is here, but I listen to his rhythmical murmuring sounds. I am in my own world.

The neurologist orders an MRI scan and I'm suddenly in a white, elongated tunnel that makes loud rapping noises that startle me. Did I have a stroke? No, that only happens to old people, doesn't it? Strokes only happen to those people that have many medical problems. It doesn't happen to 43-year-old women, and it certainly can't happen to me! However, I have left-sided weakness. It is as though someone has drawn a line down the center of my body from the tip of my head to my left toe. Everything on my left side is numb, heavy, and unknown to my body. My left cheek, half my nose, left ear, left eye, my left leg and left foot are useless as if erased from my memory bank.

"The MRI confirms it. You have had a stroke," says the neurologist. Although I cannot understand everything, I know the word. I begin to cry at the confirmation of this dreaded nightmare. As I cry, I feel the tears only on my right cheek. He speaks slowly while carefully drawing a diagram as he continues, "A blood clot or several small clots called shower emboli broke off from somewhere in your body, went through the heart, and were pumped to your brain stem and cerebellum." I know that what he says is true. I have to face it. I begin swearing in words only I can understand. All the feelings of anger, frustration, depression, and hatred tumble and stumble through my mind as I see my prior life melt away. I want to scream but nothing comes from my mouth. Am I going to be able to recover? Will this happen again? If it does, will I be strong enough to fight it? Am I going to die? I am petrified with fear.

June 10   I am extremely dizzy. It is as though I am on a ride at the fair and cannot get off. I try to get up to use the bathroom, but the nurse notices me sitting on the side of the bed and tells me I need to use a walker, an aluminum frame that steadies my gait. My left leg cannot support my body and my left arm is heavy and numb. The staff reminds me not to get out of bed by myself, but I forget. They lift the railings on the side of my bed as a gentle reminder. I sleep.

I awake to four faces staring at me and I panic. What? Who? Where am I? They are smiling, but it is a forced grin and I can see their concern through their pale faces. I feel like Dorothy in *The Wizard of Oz* as I await someone saying, "Remember me?"

I remember their names and birth dates but not their ages or any context of a past or present relationship to them. I want to jump out of my skin and my brain does not register the words they say. Something is dreadfully wrong! Someone says something about leaving the baby back at home. What baby? They laugh. I try to remember a baby. I see the hospital room and outside the window the aqua sky holds cotton-ball clouds. I fall back to sleep. Was it a dream? Was it Larry, Paige, Mark, and Betty Rae?

I ESCAPE IN MY DREAMS NOW. I had a dream that I saw God. He was a wonderful-looking young man with blond hair, very muscular with a beautiful tan. I said, "God, look at me! Look at my left arm and my left leg! Can you help me?" He showed me a large room that contained shelves of legs and arms. I was not frightened. He said, "Why don't you go in and get a new one?" I went into the room. I unscrewed my arm and put a new one on. Then I did the same for my leg. I felt wonderful. I smiled, hugged him, and kissed him. I said, "Can I use these? Are these okay?" He looked at me, laughed, and said, "It really doesn't matter, they're just parts. They are not who you are." I awoke in pain, yet thinking of that dream. We will get through this together, God and me.

The neurologist explains that the brain stem controls eye movement, hearing, and balance. The parts of the brain that lead to the central nervous system, controlling strength and sensation, have also been damaged. The thalamus in the middle of the brain has been affected by this insult. There is swelling and we must wait until this subsides to find out what paralysis, permanent or temporary, I will have. The cerebellum has also been damaged. This part of the brain controls coordination and how smoothly muscles work. As the doctor talks, he draws another diagram on the chalkboard close to the bed. I strain to make out the figure, staring at it in disbelief. I can only hear part of what he is saying. There is a constant buzzing in my ear. I am dizzy from trying to concentrate. When I attempt to talk, I do not make sense. I can't remember the words. Where did my language go? I don't know people and I am constantly frightened. Don't talk to me! Go away! Somehow my mind is transferred to the neurologist as he pivots and leaves the room. Bang! Boom! Thoughts cross my mind and go without a conclusion.

I make guttural noises for the nurse and point with my right hand to a rash that has spread over my entire body. I lie on stiff, bleached sheets as the nurse administers cream to stop the itching. I feel as though I am in a riverbank camping area during the mosquito season.

Without warning, I have a grand mal seizure. I begin having frequent headaches after this episode. The regimes of medications begin: an antiseizure medication, blood thinner, and a medication for pain.

The next day it is even more difficult for me to understand speech or to communicate. When I awake, I am still very drowsy. I can't make it through this! I just want to sleep. When people speak to me, it sounds as if they are speaking a foreign language. When I try to speak, the words don't come out right. However, when I close my eyes I can see myself inside, totally whole. I am trapped in a body that will not function.

What Did the Doctors Know—and When?

CLEO HAD CLEARLY HAD a severe stroke. Although *stroke* is a common term recognized readily by most people, few know much about what the word really means. The term was very familiar to Cleo from her background as a nurse. She too, however, realized that she knew very little about stroke.

In her recollections preceding her first journal entries, Cleo gives some hint of two factors that, in retrospect, may have represented risks for stroke. She admits to being heavier than she should have been for her height, and she hints at the many pressures and stresses in her life and of working exceptionally hard. In addition, she has had various temporary symptoms of brain dysfunction that should have warned her and her doctors that she was headed for a stroke. Strokes, like most other illnesses, seldom occur out of the blue without cause or warnings. Some hereditary conditions, other medical and cardiovascular disorders, and some habits, such as smoking, overeating, drinking too much, and lack of exercise, make some people more likely to have strokes than those who do not have these conditions or habits. These predisposing conditions are often referred to as stroke risk factors.

Some conditions that predispose individuals to stroke are already present in infancy and early childhood. Strokes occur at all ages including infancy, childhood, adolescence, and early adulthood. During the last thirty-five years, I have taken care of more than 500 patients who have had strokes before their fortieth birthday. Although it is true that strokes are much more common in those over 65, they also often occur in younger individuals. Cleo sounds a warning to all her readers. Stroke can happen to you as well as to your parents and grandparents. Beware.

Some risk factors are hereditary and cannot be helped. We cannot select our parents and our genes. Other risk factors can be eliminated or at least modified if individuals become aware of them early enough. I will say more about the various risk factors later in the book. Clearly, it is very important for all of us to become aware of our own risk factors. Prevention of stroke is far better than treatment after a stroke has already occurred.

The importance of heredity was brought home to me when I heard a schoolteacher tell of an exercise she performed with her students. The school was in the southern United States, and the students were 12 or 13 years old and were evenly mixed racially. The teacher asked her students about illnesses in their families. Most of the children knew little about the various medical conditions in their families, since parents often do not want to frighten children about their own medical problems. The teacher gave the children a homework project that consisted of asking their parents and other relatives about family medical histories. Some children had a strong family history of diabetes, overweight, high blood pressure, high blood cholesterol, strokes, or heart attacks. When doctors checked the children, many also had abnormalities that could be predicted from their family histories. Children whose parents had high blood pressure often had blood pressure that was higher than normal. Children whose parents were overweight often weighed much more than they should. Children who did not have important cardiovascular disease or risk factors in their families had a very low frequency themselves of important risk factors. Since risk factors vary so much among people, it is crucial for each individual to learn about his or her own risk factors as early as possible and to have these factors monitored all through life.

The medical problems that cause strokes often also cause the temporary symptoms Cleo mentions experiencing earlier. Doctors refer to these temporary symptoms as TIAs (transient ischemic attacks), or sometimes as "brain attacks." These symptoms are the same as those that occur during a stroke except that they are temporary, often lasting only a few minutes. When doctors recognize that a patient's symptoms are TIAs, they can often diagnose the medical abnormality and give treatments that can prevent a stroke from developing.

A TIA signals trouble in the arteries that bring the brain oxygen, sugar, and other nutrients necessary for the brain's survival. TIAs are caused by temporary blockage of an artery by a passing blood clot or a temporary inadequacy of blood flow through a narrowed artery. When the lack of blood flow is brief or relatively minor, there may be temporary loss of function that returns to normal when blood flow is restored. These brief, temporary attacks indicate that something is wrong with the system and so warn of the possibility of a stroke. Finding that abnormality and fixing it—for example, clearing the artery with a surgical procedure—can prevent a stroke.

Doctors, Patients, and Communication

CLEO VISITED DIFFERENT DOCTORS, had tests, and even had thyroid surgery, yet she kept getting worse. What went wrong? For some reason, although it now seems crystal clear to me as a reader of Cleo's descriptions, the doctors she saw at the time failed to recognize that a potentially serious brain disorder was present. Instead they tested mostly her heart and endocrine glands. Medical knowledge has grown so quickly that not even the very best primary care doctors can always keep up with the latest knowledge about all body systems.

Could part of the problem have been with Cleo? She tells us that during this time she was not thinking and communicating as well as she had in the past. Her mind wasn't working quite right. Perhaps she wasn't able to describe and communicate her symptoms clearly enough. This might seem inconceivable for someone who, like Cleo, has had medical training. But when they are sick, doctors, nurses, and other health care workers are often no better at a self-diagnosis than untrained people. It is also very human and natural for individuals not to want to believe that they are seriously ill. To most of us before we become sick, illness is something that someone else gets—not me. Denial of illness is common, and some people understate their symptoms to doctors. Other patients tend to overstate and exaggerate symptoms. Your own regular physician knows what you are usually like, but it is often difficult for a new doctor to judge the importance and seriousness of a problem during initial visits. Also, unfortunately, many physicians know very little neurology and even less about brain function.

Nor is it easy to be a patient. All doctors will tell you that the key information is your history—what you tell the doctor—upon which they base their initial diagnosis and their differential diagnosis of the likely things wrong with you.

The physical examination is important but less productive than the history described by the patient. Tests are selected on the basis of the history and examination, and tests are useful only if the right ones are done. Yet despite the critical importance of their medical histories, how many patients take time before a visit to organize their thoughts and review in their minds (or on paper) the most important symptoms and the chronology of events? The next time you feel ill enough to go to your doctor, think about the time you spend on less urgent activities, such as shopping and "to do" lists, and write your symptoms down.

Some of the symptoms Cleo asked the doctor to shed light on—tiredness, numbness of the limbs, joint pain, and change in personality—are not very specific and are common to many medical and psychological disorders not related to disease within the brain. Today she is not really sure if she told her doctors (or emphasized enough) the other things mentioned here—bad headaches, weakness on one side of the body, intermittent left-hand numbness and awkward use, unexpected difficulty understanding words that she read aloud, unsteadiness on her feet, wrong answers, slow speech, and so on. Knowing that she wrote so thoroughly about these observations months later makes me certain that she was able to describe her symptoms then as well as she does now. If she had, it is hard for me to understand how the doctors did not quickly identify her brain as the sick organ. On the other hand, maybe she really didn't want to think that she was sick and so didn't report everything. Maybe Cleo's brain dysfunctions at the time limited her ability to understand and clearly explain her symptoms. Maybe the doctors just weren't listening or maybe they didn't know enough about the brain. Some of Cleo's symptoms were only temporary, and perhaps their coming and going made her think that they were not as important as symptoms that come and remain. Temporary symptoms are very common before a stroke and warn that something is amiss.

The explanation probably lies in a combination of these possibilities.

The Problem of Unclear Symptoms

Cleo records an ominous accumulation of symptoms: headache, weakness on one side of the body, difficulty with visual depth perception. Symptoms of stroke are very diverse and more complex than those of diseases of other body organs. This is because the brain's job assignments are quite localized. In other organs in the body—the liver, lungs, and kidneys—all the cells and tissues are doing pretty much the same thing; one part of these organs looks the same as another and works the same as another. Not so in the brain.

The various brain regions look different, act differently, and contain different chemical substances to transfer information from one nerve cell to another. For instance, moving a limb, feeling a coin in a hand, seeing, talking, reading, smelling, walking, and hearing, to mention only some key body activities, are initiated in very different parts of the brain. The sides of the body, and even the tasks of the individual limbs, are connected to different parts of the brain. The left side of the brain generally controls the activities of the limbs on the right side and is involved in perception and analysis of various signals from the senses (feeling, sound, vision) that come from the right side of the body and the space around the right side of the body. The right side of the brain takes care of the same functions on the left side of the body.

Many psychological and general medical problems give general symptoms of brain dysfunction, such as feeling low, tired, sleepy, confused, generally weak, and so forth. When Cleo went to the doctor in May, the combination of weakness on one side of her body, difficulty focusing her eyes and seeing, and her headache clearly pointed to the presence of problems within specific areas of the brain. Depression, which she says her general practitioner suggested might be the cause, does not affect only one side of the body. The most common disorder that would cause such a group of symptoms, affecting one side of the body and one part of the brain, is stroke.

The Two Types of Stroke

THE TERM *stroke* describes brain injury caused by an abnormality of the blood supply to a part of the brain. The word is derived from the fact that most sufferers are struck suddenly by the vascular abnormality. Abnormalities of brain function begin quickly, sometimes within an instant. *Stroke* is a very broad term that describes several different types of vascular disease involving the blood vessels that supply the brain with needed nourishment and fuel. Since treatment depends on the type of stroke and the blood vessels involved, it is very important for the doctor to determine precisely what caused the vascular and brain injury and where the abnormalities are located.

Strokes fall into two very broad groups: ischemia and hemorrhage. Cleo had the most common type of stroke—ischemia—which means a lack of blood. Hemorrhage and ischemia are polar opposites: in hemorrhage, too much blood collects inside the skull; in ischemia, there is not enough blood supply to allow survival of the affected brain tissue. About four strokes out of every five are ischemic. When a part of the brain is not getting adequate blood, it may stop performing its usual tasks. A good comparison is the fuel pump in a car. If a fuel line is blocked and you step on the gas pedal, the car will not go because of the lack of fuel. But when the fuel line opens, the car will return to its normal behavior—and the car is not necessarily damaged. When the blood supply to a part of the brain is deficient for enough time, the tissue dies. The death of tissue caused by ischemia is called infarction. With CT scans we can tell whether the brain contains a hemorrhage, which looks white on the scan, or an infarct, which shows damage as black or gray. The CT scans

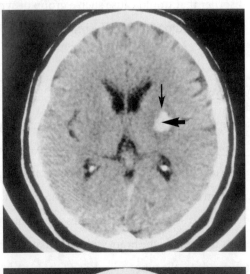

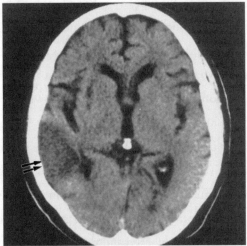

Computed tomography, or CT, *uses ordinary X rays and computers to create images through various levels of the brain. These CT scans were taken from different patients, with the scanner looking through the cerebrum, the uppermost section of the brain.*
• Top: The white region (large black arrow) represents a hemorrhage. Around the hemorrhage is a small region in which tissue is a darker gray (small black arrow) than the surrounding brain. This is edema, water in the tissue, around the blood.
• Bottom: The dark area (two small black arrows at the left) represents a brain infarct, an area of dead or injured brain tissue resulting from ischemia.

on the facing page show a small hemorrhage and a brain infarct. Often CT scans taken soon after ischemias are normal, as in Cleo's case.

There are three major categories of brain ischemia: thrombosis, embolism, and systemic hypoperfusion. Each indicates a different reason for decreased blood flow. I find these terms easiest to explain by comparing them with house plumbing. Suppose that one day you turn on the faucet in the bathroom on the second floor and no water comes out, or it comes out in an inadequate drip. The problem could be a local one, such as rust buildup in the pipe leading to that sink. This is analogous to *thrombosis,* a term used to describe a local process occurring in one blood vessel region. Atherosclerosis or another disease narrows the artery. When the artery becomes very narrow, the resulting change in blood flow causes blood to clot, resulting in total occlusion of the artery. Clearly this is a local problem in one pipe; a plumber would attempt to fix the blocked pipe. Similarly, physicians treat a narrowed (stenosed) or occluded artery by trying to open it or by creating a detour around it.

But a blocked pipe to a second-floor sink could also be caused by debris in the water system that came to rest in that pipe, rather than by a local problem that began within the pipe. When particles break loose and block a distant artery, we call it an embolism. (The place where the material originates is called the donor site; the receiving artery is the recipient site; and the material is called an embolus.) An artery within the head can be blocked by a blood clot or other particles that break loose from the heart, from the aorta (the major artery leading away from the heart), or from one of the major arteries in the neck or head. An embolism was the cause of Cleo's stroke.

Suppose instead that the plumber finds that the water did not flow normally in your second-floor sink because the water pressure in your house is intermittently low and flow to all sinks and showers is faulty due to a leak in the water tank or low water pressure in the entire house plumbing. This situation is like systemic hypoperfusion: there is no local problem with the pipe

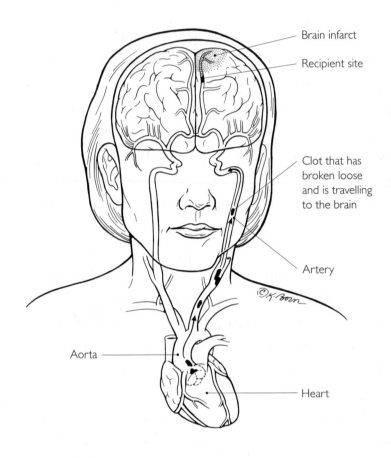

Brain infarct

Recipient site

Clot that has
broken loose
and is travelling
to the brain

Artery

Aorta

Heart

Brain embolism: *The heart, aorta, and major arteries in the neck and head are the usual places where clots or other particles break loose and travel to arteries in the brain, causing a brain infarct.*

to one sink, but instead a general circulatory problem. Ischemia can be caused by inadequate pumping of blood from the heart or a low volume of blood or fluid in the body.

LIKE STROKES DUE TO ISCHEMIA, those caused by hemorrhage— the second major type—can occur for any of several reasons. *Hemorrhage* refers to bleeding inside the skull, either into the brain or into the fluid surrounding the brain. Bleeding into the brain is called intracerebral hemorrhage. It damages the brain by tearing and disconnecting vital nerve centers. The bleeding is due to the rupture of small blood vessels, arterioles and capillaries, within the brain substance. This kind of hemorrhage may be caused by head trauma, high blood pressure, an abnormality of blood clotting that causes excess bleeding, or a congenital blood vessel abnormality (vascular malformation).

The blood oozes into the brain under pressure and forms a localized, often round or elliptical blood collection, called a hematoma, that separates normal brain structures and interrupts nerve connections called tracts. It also exerts pressure on brain regions next to the collection of blood and can injure these tissues. Hematomas affect abilities or parts of the body related to the brain area damaged by the blood collection. For example, if the bleeding is into the left cerebral hemisphere, the patient often has weakness and loss of feeling in the right limbs and a loss of normal speech. A hemorrhage into the cerebellum will cause dizziness and a loss of balance. The particular symptoms are associated with the region of bleeding and are described as *focal*—that is, they are related to dysfunction of only one brain region.

Bleeding into the fluid that surrounds and bathes the brain is called subarachnoid hemorrhage because blood collects between a spiderweb-like membrane called the arachnoid and the pia mater, tissue that fits over the brain like a piece of Saran Wrap. Subarachnoid hemorrhage is usually caused by the rupture of a weakened artery whose wall is ballooned outward. This condition is called an aneurysm. The artery breaks, spilling blood instantly into the cerebrospinal fluid that circulates around the brain and spinal cord. The sudden release of blood under high pressure increases the pressure inside the skull and causes severe sudden-onset headache, often with vomiting. This sudden increase in pressure causes a lapse in brain function, so that the patient may stare, become confused, unable to remember, or experience a sudden change in equilibrium. Most often the symptoms in patients with subarach-noid hemorrhage show up as diffuse abnormalities of brain function because usually there is no bleeding into the brain. The skull forms a closed hard sphere, a fortress around the brain and its surrounding membranes. With subarachnoid hemorrhage, the fortress can become a prison that keeps the blood and pressure within the skull. In this closed system, the bleeding causes pressure to build up quickly and strangle normal tissues by compressing them.

CHAPTER 2

Stroke: The Battleground

ON JUNE 9, Cleo had her stroke, but in her journal it seems that she did not fully appreciate or understand what was wrong or what had happened to her. She describes a variety of symptoms, including confusion, difficulty communicating, problems with understanding speech and events, and loss of normal vision. These functions all relate to the cerebrum—the so-called thinking portion of the brain. But in the days immediately following the stroke, we see her become aware she is terrifyingly impaired, evidence that she has suffered widespread damage in her brain.

Some of Cleo's symptoms the day after her stroke—dizziness, vomiting, ear buzzing, uncoordination of her right limbs, and difficulty walking—were caused by ischemia to her brain stem and cerebellum. The photographs on pages 34 and 35 are pictures from Cleo's MRI. With them are drawings identifying the various brain structures her stroke damaged. Cleo had infarcts in her cerebellum, in the right temporal and occipital lobes, and in the right thalamus—a severe brain attack involving her powers of thought, emotion, vision, language, and movement.

How the Brain Works

THE BRAIN IS without question the most important and interesting organ in the human body. It is responsible for our movements, feelings, moods, thoughts, and perceptions, and it allows our unique personal characteristics, our abilities and failings, our intelligence, and our personalities. In short, our brains make us what we are. The brain is also the most intricate and complex computer system known.

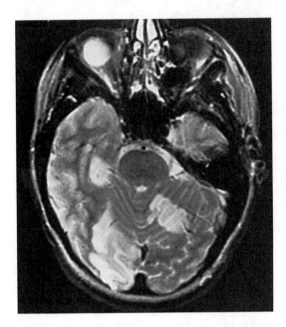

This magnetic resonance imaging (MRI) scan was taken through Cleo's brain stem, cerebellum, and occipital and temporal lobes. The scan reveals infarcts (the white areas) to her right temporal and occipital lobes and in her cerebellum.

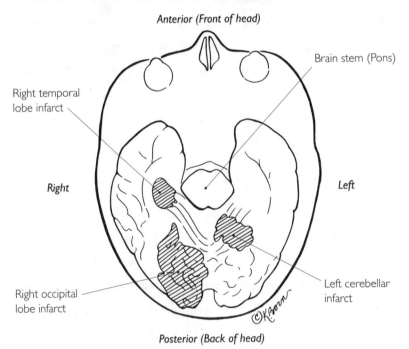

Areas of Cleo's brain damage shown in the MRI.

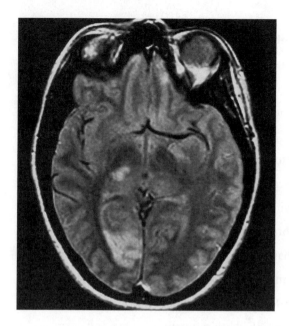

A frame of Cleo's MRI taken a bit above the one on the facing page. Here the scanner discloses further damage in her right occipital lobe and an infarct in her right thalamus.

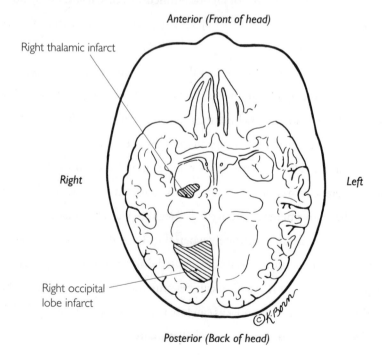

Anterior (Front of head)

Right thalamic infarct

Right

Left

Right occipital
lobe infarct

©K.Bonn

Posterior (Back of head)

Areas of Cleo's brain damage shown in the MRI

The brain stem and cerebellum

The drawings on the next two pages show the human brain stem and cerebellum. The brain stem, a small but critical structure located in the back of the head under the cerebrum, connects the spinal cord with the thalamus and cerebrum.

The brain stem is vital for five reasons: 1) it contains nerve cells involved with the movements and senses of the head and face, 2) it acts as a passageway for information traveling to and from the cerebrum, 3) it acts as a relay station for information coming to and from the cerebellum, 4) it contains nerve cells and pathways that maintain consciousness and alertness and relate to sleep and wakefulness, and 5) it houses nerve cells that control automatic body functions such as breathing, heart rate, and blood pressure. Destruction of the brain stem leads to loss of all brain functions, coma, and death, and a stroke in the brain stem that does not kill can lead to paralysis of limb, face, and mouth muscles.

The cerebellum, whose name literally means "little cerebrum," or "little brain," looks like a small walnut attached to the brain stem. It is located far back in the head, below and behind the cerebrum, which dwarfs it in size. The cerebellum and brain stem communicate by way of three nerve bridges— from the brain stem's lower (medulla), middle (pons), and upper (midbrain) areas. The cerebellum contains nerve cells and pathways that are important in ensuring that movements of the arms and legs are well coordinated, that the eyes move smoothly and accurately when they look at objects, and that the lips, mouth, and tongue are well coordinated during speech. When a stroke damages the cerebellum, walking can become unbalanced, talking can become slurred, and the arms and legs can lose their normal coordination. Cleo's loss of balance was caused by her cerebellar infarct.

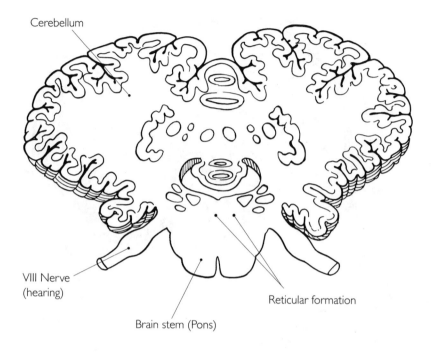

Cerebellum

VIII Nerve
(hearing)

Brain stem (Pons)

Reticular formation

A look at the brain stem and cerebellum. *This drawing represents a cut crosswise through the brainstem at the level of the pons and the cerebellum.*

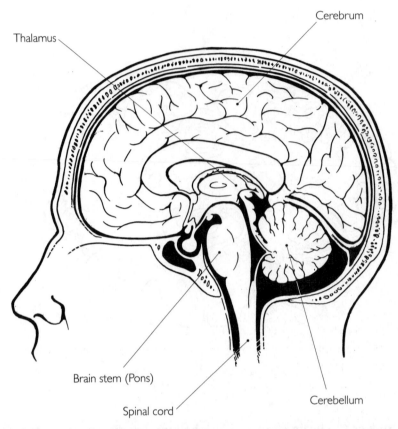

Cerebrum

Thalamus

Brain stem (Pons)

Spinal cord

Cerebellum

The brain stem is an upward continuation of the spinal cord and is connected to the cerebellum by three bridgelike structures. Long nerve pathways called motor tracts lead from the cerebrum to the base of the brain stem and continue into the spinal cord delivering messages to move the body's limbs. The main tracts carrying sensations from the body to the brain also travel through peripheral zones of the brain stem on their way up to the thalamus and cerebrum. Within the brain stem, but not visible in these drawings, lie many important nerve cells, or neurons. Mostly toward the back are nerve-cell groups, called nuclei. Some control eye, tongue, face, and neck movements; others receive sensory input related to sounds, movement in space, touch, pain, temperature, and pressure felt in the face, throat, mouth, nose, and ears. On each side, near the middle of the brain stem, are small neurons, called the reticular activating system (shown on p. 37), that maintain alertness and an awake, energized state and that help control the sleep-wake cycle. If the reticular activating system on both sides of the brain stem is injured, coma develops.

The outside of the brain and how some brain tasks are organized

The drawing on page 40 shows the outside of the cerebrum from the left and from the top. Farther back and below, the cerebrum covers the cerebellum and brain stem (not visible in this drawing), much like the cap hovers over the rest of a mushroom. The largest portion of the brain, the cerebrum's two sides are called the left and right cerebral hemispheres. These hemispheres are divided into sections called lobes.

By far the most familiar feature of the brain is its surface, called the cerebral cortex. The cortex is made up of folded, raised strips of brain tissue, called gyri, with valleys or clefts, called sulci, between them. The parts of the brain in front of the central sulcus are called the frontal lobes, which are mostly related to action and movement, so-called motor functions. Behind the central sulcus are the parietal, temporal, and occipital lobes, which are more related to the senses—our perceptions of various stimuli in the environment, such as sound, vision, and touch.

Visual deciphering mostly takes place in the occipital lobes; the somatosensory (from *somato*, a Greek term that refers to the body) cortical areas are located in the parietal lobes; and the auditory regions are located in the temporal lobes. The brain also processes smell and taste in the temporal lobes.

Each sense handles input in a similar way. Simple sensory signals first go to special regions called primary cortical regions. Scientists have named these regions for the stimulus they respond to. For example, the primary cortex for vision is called V1, the primary cortex for sound (auditory input) is called A1, and touch and other sensations from the body (somatosensory input) are first received at S1. The signals that reach these primary areas are usually quite simple: V1 detects light and lines, A1 simple sounds and noises, and S1 simple touches and pressures on the skin.

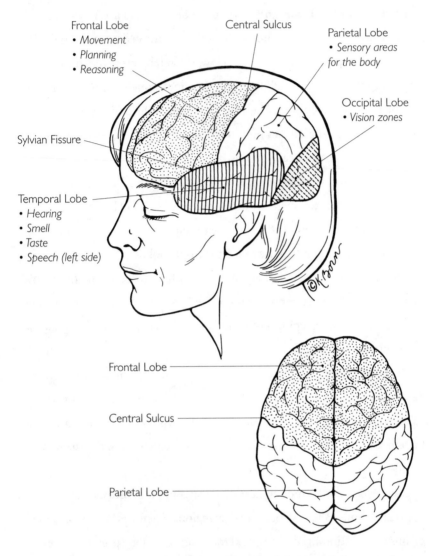

The outside of the brain, here shown from the left side and the top, is the most familiar view. Visible from this perspective are the "lobes," the regions involved in the myriad aspects of human thought and behavior. The frontal lobe (speckled area), lying in front of the central sulcus, is a key region for movement, planning, and reasoning. Behind the central sulcus, the parietal lobe (unshaded), is important for processing sensations from the body. The temporal lobes (marked with vertical lines in the side view) lie behind the temples and process hearing, smell, taste, and, on the left side, speech. The occipital lobe (crosshatched in the side view) is responsible for processing vision.

Next to the primary sensory regions are secondary sensory cortical regions, where more elaborate sensory information is analyzed and processed. For vision, the information processed might be boxes, circles, forms, and so on. Next to the secondary cortical regions are tertiary zones that process even more complex input, such as faces, animals, and scenes in the visual sphere, and words and musical phrases in the auditory sphere.

Each side of the brain relates to the opposite side of the body and the space around the body. The right visual area of the brain relates to vision on the left. Cleo's infarct in the right occipital lobe interrupted visual input so that she lost the ability to clearly see objects and people to her left.

By the sixth day after her stroke, Cleo had become aware that things were not working as before. She had lost the feeling in the left side of her body, and although she hadn't registered this yet, she also could not see objects to the left. She was mixed up and had difficulty with language and memory. She was dizzy, nauseated, and uncoordinated when she walked. As time goes on, she will be able to describe her losses in more detail.

The four drawings on pages 43–46 tell the story of how communications take place between the brain and the body. The drawing on page 43 shows the inside of the brain. Nerve cells are the brain's gray matter. They are located in the cerebral cortex, the basal ganglia, and the thalamus. A stroke affecting the basal ganglia will interfere with the action of the limbs and body. Strokes (like Cleo's) affecting the thalamus interrupt the signals from the senses to the cerebral cortex. All fibers that travel to and from the cortex are wrapped in myelin, a fatty coating that allows faster transmission of nerve impulses through the fibers, much as insulation does around ordinary electric wires. Myelin gives the brain a white color, which is why the regions through which connecting nerve messages travel are called white matter.

Strokes involving the white matter disrupt communications between nerve cells. Depending on what pathway a stroke damages, the results can be as different as loss of movement, vision loss, or language impairment.

The drawings and captions on pages 44 and 45 tell about the motor pathway. The brain's main motor pathway originates from the primary motor cortex within the gyrus (the "precentral gyrus") just in front of the central sulcus. Damage to the motor cortex or along this pathway at any point leads to loss of voluntary motor control of any parts of the body below the interruption. The weakness of Cleo's left arm and leg were related to ischemia of this tract in the midbrain, or pons. The drawing on page 45 shows the various pathways within the motor system. Fibers travel from the cerebral cortex through the internal capsule, brain stem, and spinal cord to reach the nerves and muscles in the arms and legs.

And finally, in contrast to the motor system, which passes information from the cortex to the body, are the sensory systems shown on page 46. These systems bring information to the cerebral cortex so that we become aware of what we see, hear, feel, taste, and smell. Other input tells us about movement within the environment and the relation of our bodies to the outside. In general, these systems have a common pattern. Sensory receptors within the eyes, ears, nose, tongue, and skin first receive input from the environment. The somatosensory receptors within the skin and the bone and joints send information to neurons within nerve clusters that lie outside the spinal cord. These neurons send information toward the brain about touch, pain, heat, cold, pressure, and the location of the limbs in space. They also send information locally to the motor nerve cells so that automatic reactions (reflexes) such as limb withdrawal can occur without need for the information to pass through the brain.

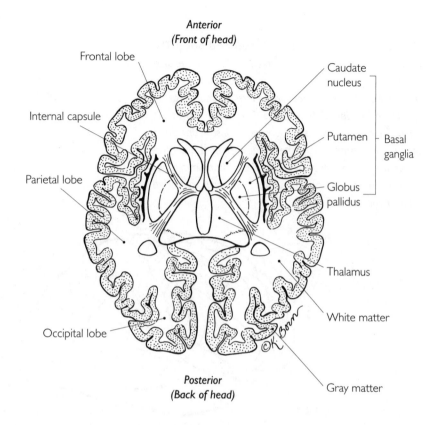

Communication in the brain. *This drawing of the brain as if cut horizontally across the middle shows key regions for communication among brain cells. The cerebral cortex is the speckled ribbon along the outside circumference. Here are found many of the nerve cells that relate to motor, sensory, cognitive, and behavioral functions. These neurons give rise to nerve fibers that communicate with other parts of the cortex. Fibers travel from the cortical neurons downward toward groups of neurons, called the basal ganglia, embedded within the brain (labeled caudate nucleus, globus pallidus, and putamen) and to neurons located in the brain stem, cerebellum, and spinal cord. The internal capsule, lying between the caudate nucleus and the rest of the basal ganglia, contains fibers controlling limb movement. Many fibers also travel upward from the spinal cord, brain stem, cerebellum, and basal ganglia toward the thalamus and the cerebral cortex—mostly to relay information about what is happening in the environment and in the body.*

Communication between brain and body. *The central sulcus is a kind of Mason-Dixon Line: on the side toward the front of the head is a raised strip of brain tissue called the precentral gyrus. Here the motor cortex lies, like a headband, from one side of the brain to the other. The motor cortex is responsible for instructing parts of the body to move. On the other side of the central sulcus, the postcentral gyrus seems like a mirror image, but here resides the sensory cortex. Its business is to receive information from the body.*

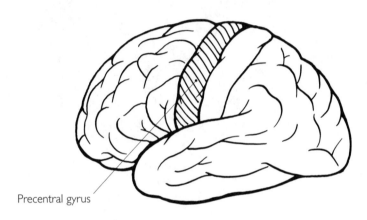

Precentral gyrus

The motor pathway from brain to body (facing page). *Sketches of body parts at the top of this drawing show the locations of motor cortex neurons that produce movement of particular body parts. Fibers from all the motor cortex neurons form a tract—the pyramidal tract—that descends within the white matter, then travels through the front portion of the internal capsule. From there, the tract continues down to the base of the brain stem, where some fibers leave the main path to connect with nerve cell groups in the brain stem that control movements of the eyes, face, jaw, and tongue. The main pyramidal tract goes on, however, descending into the spinal cord to connect with motor neurons of the spinal cord that control the muscles of the trunk and limbs. In the upper neck (cervical portion of the spinal cord), more fibers leave the tract to take control of intentional movement of the upper limbs. Lower down, in the lumbar, or lower back, region, still more fibers leave the tract to control the lower limbs. The rest of the pyramidal tract fibers go all the way to the lowest end of the spinal cord (the sacral, or tail, region) and connect with neurons that help control pelvic and genital muscles relating to urination, defecation, and sexual function.*

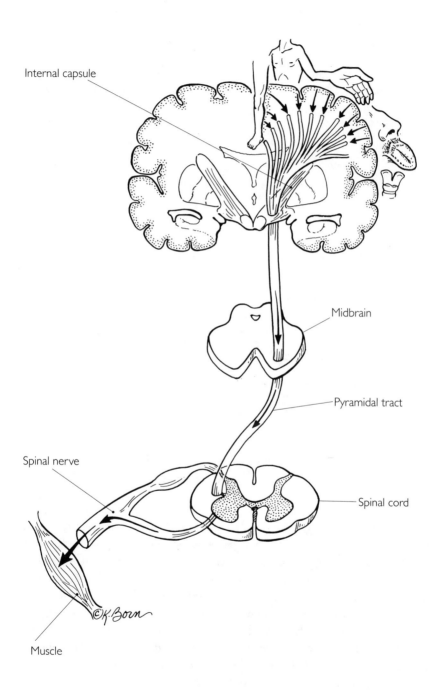

Internal capsule

Midbrain

Pyramidal tract

Spinal nerve

Spinal cord

Muscle

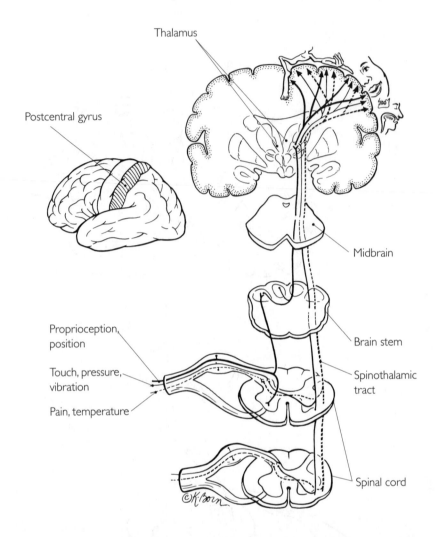

The sensory pathway from body to brain. *The body parts depicted at the top of this drawing mark the locations in the sensory cortex where specialized neurons receive information from the body. The tracts for pain and temperature sensation and limb position cross to the opposite side of the spinal cord and then ascend toward a very large, centrally located structure called the thalamus, which sits on top of the brain stem at the foot of the cerebral hemispheres and is the main way station or relay for sensory input to the brain. The sensory information is then transmitted from the thalamic nuclei to a somatosensory area in the parietal lobe located in and around the gyrus just behind the central sulcus (postcentral gyrus). When the information reaches our brain, we become aware of sensory signals on the opposite side of our body.*

In the brain, the motor and sensory pathways cross over to end up opposite the side of the body from where they started. As a result, damage in one side of the brain will cause symptoms on the opposite side of the body.

The visual and auditory systems have a similar pattern of relay of important information to the brain. Light and sound are first perceived in nerve cells in the eye (retina) and inner ear (cochlea) and then go through nerve bundles called tracts toward special areas in the thalamus, from where, as with other sensory systems, the information is relayed to specialized regions in the brain.

The drawing on page 48 shows the way vision is handled by the eye and brain. Input from each eye travels through the optic nerve located behind that eye. When an injury occurs to your right eye or right optic nerve, you can't see out of the right eye, but you see normally using the left eye. However, injury to nerve tissue deeper in the optic system can take away pieces of vision entirely. Here is how it works: Some fibers within each optic nerve cross to the other side of the brain so that each eye's visual signals for the right side of the world now travel through a fiber bundle at the base of the left side of the brain. The fiber bundle is called the optic tract. If you draw a line right in the middle of your vision, everything on the right is being handled in your left optic tract. Similarly, the right optic tract contains information about everything on the left. After passing through the thalamus, the optic fibers finally end up in the visual cortex. The left visual cortex receives information from the right side of the visual field, and the right visual cortex receives information from the left visual space.

If you have an injury to the right optic tract or the right visual pathway within the brain, you cannot see to the left (a left visual field defect) using *either* eye. Cleo's left visual field loss was due to her right occipital lobe infarct. The idea that the brain is organized in relation to the sides of visual space is very hard to grasp, since in everyday life we tend to think of vision solely in relation to the individual eyes.

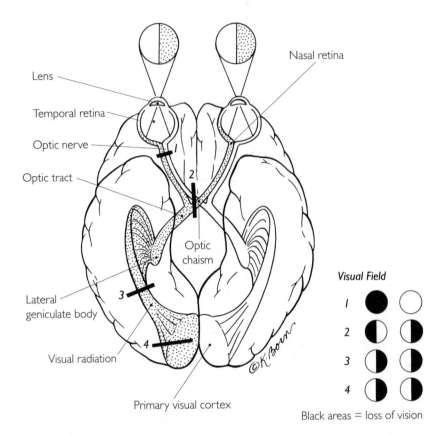

Nasal retina

Lens

Temporal retina

Optic nerve

Optic tract

Optic chaism

Lateral geniculate body

Visual radiation

Primary visual cortex

©K.Born

Visual Field

1

2

3

4

Black areas = loss of vision

When vision is damaged. *Vision loss can be caused by stroke without the slightest damage occurring to the eye itself. Here's why: The visual pathways extend from the eyes to the visual areas in the occipital lobes. The outside, or temporal, part of the retina of each eye views the middle portion of scene in front of it, while the inner (nasal) part of the retina takes in the outer parts of the scene. Visual information goes from the retina of each eye through the optic nerve, ending up in the visual cortex, where the scene's content is sorted out. In this drawing, numbered bars mark the locations of hypothetical damage along the visual nerve pathways. The numbers correspond to circles at the right showing what parts of the scene are lost if a person has that particular damage. In scenario 1, an optic nerve is damaged, and the person cannot see from that eye. In 2, damage occurs in the optic chiasm where the optic nerves cross. That interference with visual fibers from each optic nerve causes the loss of side (temporal) vision in each eye, resulting in so-called tunnel vision. In both 3 and 4, the damage affects the fibers after they have left the optic chiasm. Having crossed over, those fibers are handling information about the scene on the opposite side. Thus, at this point, damage on the left side blacks out vision of the right side of the scene.*

SOUND INPUT REACHES the brain through pathways that eventually change sides, too. The right temporal lobe receives sound information coming from the left, and the left temporal lobe relates to sounds coming from the right. If an abnormality develops in one ear or one auditory nerve, then you have difficulty hearing sounds with that ear. But farther into the brain's pathways for hearing, an injury to one of the pair of structures called the medial geniculate bodies or to one of the temporal lobes affects all sound coming from the opposite side no matter which ear hears the sounds. Cleo describes some hearing difficulty that might have been caused by the injury to her right temporal lobe.

Cleo had areas of damage in particular right-brain sensory pathways. Her MRI showed an infarction in her right thalamus. The damage to her right thalamus and to the fibers from the thalamus to her somatosensory cortex in the right parietal lobe caused the severe loss of all sensation in her left face, arm, leg, and trunk. In her right occipital lobe, the damage to the visual pathways and visual cortex caused a loss of vision related to the left side of her environment.

Damage to the thalamus also explains some of Cleo's confusion and difficulty remembering details. The thalamus connects to regions of the cerebral cortex on each side that have specialized abilities for memory, language, and other functions that relate to thinking and behavior.

How the brain puts it all together

How does the brain put together the separate motor systems, vision, hearing, feeling, speech, and memory? One important theme of brain function is the interrelationship between motor and sensory activities. To use vision as an example, look at the scene shown on page 50.

At first glance, most people see one key part of the scene, such as one child playing. This part is noticed by the visual cortex within the occipital lobes. But the initial visual information generates questions that you try to answer

A scene shown to a patient *who looks at it for about twenty seconds can reveal the existence of problems in several brain areas: the visual cortex, which should register key parts of the scene; the frontal lobes, which should search the picture; and memory and language areas, which should work together to draw meaning from the picture. Cleo would have had great difficulty with this simple test in the early days after her stroke; she frequently mentions trouble making sense of things she saw and heard.*

by looking further at the picture. Eye-movement searches are generated by gaze centers in the frontal lobes. Looking leads to more seeing, which in turn raises further questions and stimulates more looking. You gradually get more and more information about the picture: you may decide what you think are the ages and sex of the players, the game they are playing, the nature and location of the Victorian house. "Looking" (a motor activity) and "seeing" (a sensory operation) work together to allow you to get the most information from the picture.

Of course, the information you get from a picture depends on your intelligence and experience. You use your memory zones and your language area to attach a name and a meaning to what you see. Similarly, if I blindfold you and put an object in your hand, you will feel something. Then you will think of things it might be and then touch and explore the object with your hands to detect its nature and its name. Your perception by touch takes place in the somatosensory area opposite to the hand in which the object is placed, and your hand movements are generated in the hand and arm area in the opposite frontal lobe. Similarly, if I play music, you hear it in the lateral portion of the temporal lobes, and then you tune your ears to listen for familiar features of the music. This tuning is mostly done in the frontal lobes. Of course, as in seeing and feeling, in order to recognize the music you must have heard it before. You have no way of identifying things that you have not seen, heard, or felt before; if you don't already know the name of an object I hand you or a song I play, you're out of luck.

Another important feature of how the brain assembles information is the connection between the sensory input regions and the language and memory regions. For example, suppose I show you a familiar coffee mug. You first see the cup with your eyes and your visual cortex. The visual information goes to the language area in your left cerebral hemisphere to find a name— "mug" or "cup." The information also stimulates the visual memory area in your right medial temporal lobes and you remember that a coworker gave this particular cup to you as a present and that you use it for coffee each morning at work. If you don't have any brain damage, you activate taste and smell regions in the temporal lobes, reinforcing that coffee is the stuff placed in this mug, and you activate somatosensory parietal lobe regions to recall the feel of the mug. Activating the various senses and memory and language leads to a full characterization for you of the nature of that particular object and what it means to you.

The emotional element of perceptions may also be affected by stroke. Some things that we see, hear, feel, and otherwise experience stimulate us.

These perceptions somehow make us feel fearful, uneasy, happy, angry, excited, or sad. At times we are quite aware of the emotional content of perceptions, but at other times we are not. Perceptions are mostly located within the so-called limbic areas of the temporal and frontal lobes. The right cerebral hemisphere seems to have more influence on emotive, affective content than the left cerebral hemisphere. A stroke that affects these areas may alter the accuracy of emotional perceptions or diminish or increase their strength and so change an individual's personality traits. Cleo will find herself struggling with this problem, too.

CHAPTER 3

The War of Rehabilitation

June 16, 1992 Doctors and nurses continue to stream in and out of my room. I cannot communicate with them. I cannot make my mouth form the words. The words are there in my brain but I cannot get them out. When the nurse talks to me I want to scream, "I'm here!" I am frustrated, but more than that, I am scared.

By evening, I begin to form some small words. I walk to the bathroom with the help of a walker and a nurse. I'm back! My left side is still paralyzed, my left eye and ear are not assimilating sights and sounds correctly. However, I can communicate to some extent. I am happy about the few words I can get from mind to mouth, with a small amount of the language I have taken for granted all my life.

But I feel as if I'm on an emotional rollercoaster. I am not only afraid for myself but what this vulnerability will do to the children. I panic and cry whenever the hospital personnel mention their names. No! No! Keep them away from me! The doctor will fix me up good as new and I'll be okay. Imagine what this will do to their psyches, seeing their mother unable to communicate, unable to say, "Everything will be all right. Mom is fine, don't worry." I don't want to see anyone, especially the children.

June 17, 1992 I am uneasy, yet excited. Today I am leaving our community hospital to enter a rehabilitation center at a larger facility. The therapists refer to it as "boot camp." I can't wash my hands or position myself in bed without help. The world exists only through my right eye.

Paige is here to drive me the twenty miles or so to the rehabilitation hospital. It is a gorgeous day in June, and I do not want to be in any hospital. It feels wonderful to be dressed again, to be in the fresh air and sunshine. I smile at Paige and feel a bond between us that I have never known before. I convince her to go through a fast food place for a soda. She understands my pointing, grunts, and moans. I believe I am speaking clearly, but I recognize from her keen look that she has to listen diligently for every word I say.

We travel within blocks of our home and I feel that the nightmare is over. I want her to take me home and to forget all of this as if it were a bad dream. I try to be strong. I beg to stop in for a minute to see Mark and Betty Rae. I haven't seen them for several days and my mind jumps quickly from one thought to another, never capturing an entire thought and bringing it to conclusion. I feel hypersensitive to the car ride as snapshots of still-frame pictures barrage my sight. If I close my eyes, my body seems to be moving at warp speed and I feel dizzy. When Paige makes a left-hand turn I grab the dashboard with my right hand even though I'm buckled in. I'm frightened to see Mark and Betty Rae when I seem so out of control. I want to be whole again.

"Mom, now you know we have to register by a certain time," Paige says.

The first day at the rehabilitation hospital is devastating. As we arrive at the hospital registration area, I have to use the bathroom. I tell a nurse. Paige is busy at the registration desk, and a male nurse whisks me up to my floor, wheels me to the bathroom, and proceeds to pull my pants down to my ankles. By assisting me, he strips me of my dignity. I feel so ashamed. A male nurse had never helped me before, and it is a shock when done so abruptly. Paige returns to the two-bed ward and sits in the chair next to my bed. I try to be brave and repeat to myself, "Don't you cry, Cleo. Cleo! Don't you dare cry!" Paige tries to smile as she asks if there is anything else I need. I shake my head and gather a cassette recorder, pen, and paper to make it appear as if I've got things under control. Paige kisses me on the cheek and leaves. My world no longer revolves around the family. In fact, it is just the opposite. I have become self-absorbed. I am only aware of what is happening in the present moment and space of time. The past does not exist except for familiar faces that enter my world. I am in the bed closest to the door and if I turn my entire body to the left I can see the lady in the other bed. I begin to tape a journal entry and I hear her sob.

The room has blue-green walls and a wide plaid curtain marking each person's bed and desk area. I'm on the sixth floor of the older section of the hospital, which has been converted into a rehabilitation floor.

I am introduced to some friendly people. A therapist gives me written information and rules of the unit. They are not going to do me any good. I can't read or remember worth a darn! I am told my schedule. I cannot assimilate it all.

At suppertime, the patients eat communally in a large dining hall. Five elderly men sit in wheelchairs around an oblong Formica table, awaiting their meal. Two of the men wear white cotton bibs. A towel is draped over the chest of another. One man's face constantly winces.

A gray-haired man is slumped to one side of his wheelchair and drooling. I try to eat but, in my mind, I refuse to belong to this group. Bending my head, I weep uncontrollably.

When I return to my room, a doctor informs me that I have had several strokes. Not one or two, but several. Am I going to have another one? Will I die?

I am given a suppository. The nurses seem to be preoccupied with the frequency of my bowel movements. I'm afraid I'll soil myself. By early that evening I am lying in feces.

June 18, 1992 I have been wheeled down to the physical therapy department twice today. The sweat beads up on my forehead as I strain to control my left leg. It's too heavy to lift. Why do I have to relearn to walk? Why can't I just do it?

In speech therapy I learn to make a new sound and to say a new word—"um" and "bad." I used these words frequently to describe my condition.

"How are you today, Cleo?"

"Um, bad." I respond.

My hair is pulled back in a low ponytail at the nape of my neck and fastened with a barrette that I can clasp with my right hand. I try to apply makeup but apply it to only half of my face. I don't notice that my lipstick is smeared above the lip line.

I learn to walk between the linen carts and wheelchairs that line the wide halls. The nurses are distributing medications. I begin to learn my own scheduled curative regime in preparation for the outside world.

June 18, 1992 **EVENING OR NIGHT.** I hear voices but I can't respond or understand what they are saying. My body is stiff, heavy, and uncontrollable. Lights, people, someone is yelling, quiet voices talking all around me, asleep, fighting to stay conscious, warm blankets . . . I remember the warm blankets! That night I have a second major stroke.

June 18, 1992 *[Transcribed from memory days later.]* I awake alone in a room. I can't stay awake for more than a few seconds and then I drift off again. I hear something beeping and there are bright lights above me. There is a window at the end of the room and I seem to remember seeing Larry intently watching me through it with his hand against the glass. I can't fight sleep. The large room is empty except for my bed. The side rails are up but I can't move my heavy body.

On or about June 19, 1992 I am in Cardiac Intensive Care. Doctor Kendel, my cardiologist, explains that my heart is not working normally. A hole in my heart allows blood clots to travel from the right atrium to the left atrium and pump to the brain.

Larry is by the door during the doctor's explanation. His face appears pale and expressionless, and he stands as if at parade rest in his business suit. The doctor leaves a model of the heart on my bedside table, and as I try to assimilate this new information, Larry leaves without a word. The snapshot is imprinted on my brain: the heart, and Larry leaving. How can he leave me now? Doesn't he understand the way I feel? I don't understand—is he grieving too?

It cannot be my heart. I had a coronary angiogram less than six months ago and the cardiologist said I was fine. Apparently, without one particular test it was impossible to see the defect. Not again! I had open-heart surgery when I was a child for a valve that didn't open correctly. It was called *pulmonary stenosis*. I was referred to as a "blue baby." It was repaired in 1955. I didn't have to take medication. My children's births were normal deliveries. I had had several surgeries without complications. A hole in my heart! Am I hearing right? Is there another way this could have happened? At this point, I am not sure I want to hear it.

On or about June 20–24, 1992 I am transferred out of the Intensive Care Unit to the neurological floor. I cannot speak except for the word "bad." I use it often to describe my condition and some of the procedures. I must have been there three or four days. My mind runs thoughts together and I have no idea what date, month, or time it is. Larry is here, in and out as his work schedule permits. I sleep most of the time. I try to listen, try to comprehend but my eyelids are so heavy that everything shuts away and, for a while, I am lost.

Thank God I can write a little with my right hand and read a few basic words in large print, so not all communication and comprehension is lost. When I listen to what people are saying, it sounds strange and disjointed. I need to just hold their hands and hope this frightening experience will pass. I must take my time to write, and I can read using my far right field of vision. My glasses are no good now. The doctors and speech therapists believe I am showing signs of aphasia, which means that I have difficulty with the ability to use and understand language in various forms, such as reading, speaking, or writing. Aphasia is an unconscious foe that will strike at the most unusual time and hold me captive. *[I was unaware, at this point, that memory difficulties or some type of aphasia would be my nemesis for years to come.]* I find myself mixing words or letters. Words do not come out the way I intend. It is as if my vocabulary just disintegrates and I am reduced to a game of charades. Things that should come automatically are now a slow ticker-tape process from mind to mouth. Many times I think of myself as a life-size robot that needs batteries for thinking and processing and when my batteries wear down I must rest. Aphasia is my adversary and I don't like people treating me as if I am a child. Some medical personnel speak to me in a very loud staccato voice and it frightens me. Give me time, work with me, and help me recover.

Exact dates not recorded in journal I am scheduled for an-other MRI brain scan. This procedure mandates that I go out of the hospital to another building via a stretcher and ambulance service. As I reach the door of the hospital, I feel the breeze, cool and wonderful. I smell the fresh air and feel the warmth of the sunshine against my skin. This feeling is replaced by a Velcro chin fastener and the whirring of the coffin-like MRI machine.

LATER THAT DAY, I hear the cardiologist talking to Larry about heart surgery. The doctors want to prevent further strokes by surgery or blood thinners. What if I am too weak and debilitated to survive it all?

Occupational therapists, physical therapists, and speech therapists come every day, but I am not out of bed yet. I need to get up and start moving around.

Our neighbors for the last four years are Claire and Don. Their son is Mark's age and the boys spend a lot of time together. Claire and Don visited me this evening and brought music tapes of *Phantom of the Opera* and *Les Misérables*. I seem to remember Larry and me attending one of these Andrew Lloyd Weber productions with them at the theater, box seats if I recall correctly. I remember Don likes the song about the Red and Black ... must have been *Les Misérables*. Don and Claire always look so put together and their hair appears carefully groomed with look-alike silver highlights. We have had great times together and usually get together at least once a week.

I find the music peaceful even though I can't understand all the words. I feel the need to have something soothing to listen to and the calming effect of the soft music brings me back to a place of serenity and pleasant times.

I have lost my appetite, and with it about thirty-five pounds. Most things taste like metal. However, cold things like watermelon, iced tea, Popsicles, and water taste moist. The food is always mashed up. At times, the nurses smash my pills and put them in peach baby food. It is humiliating. I can swallow. It only takes time.

The medical costs are many, many thousands of dollars. I can't do anything about it now. I know I have fine doctors.

Larry is talking about selling the house and buying a much smaller rambler-style. He sits on the bed next to me and stares out the window as if in a daze. We had purchased the house with my child care income in mind, although we qualified for the mortgage under his income alone. I hand him a signed blank check from my business account. It was the last check I would sign from that account as it only held about eight hundred dollars. We were like everyone else, living on credit from paycheck to paycheck, never realizing our house of cards could topple over. Larry may have worried about our financial future before but had rarely voiced his concern. Now, with his voice cracking and his eyes empty of comfort, I know he is serious.

I write d-i-v-o-r . . . on a piece of paper. I think that financially speaking it would be better for the kids and Larry. I can't go back to child care, and nursing doesn't seem like a probable option either. For me, employment of any type looks far away.

"Don't," he replies. I was hoping for a better response, maybe one from the ending of an old movie. But this isn't high drama, this is real.

I am full of trepidation about going home. Will it be the same? Will I be able to move and do everything as I did before the stroke? Will I be a good wife to him? Will we be able to love each other as we once did? Will I be pretty in his eyes? Will I walk without a limp and move my arm comfortably. Will I feel his kiss?

Last night I slept very soundly but awoke disoriented. I'm afraid to sleep at night, afraid of having another stroke. I can't shake off the fear.

IT IS LATE ONE EVENING, after visiting hours, when our family physician arrives. I try to say his name but all I can do is sob. Tears flow so easily now. He has been our family physician for many years.

"Well, how are you doing? I thought I'd stop in to see my favorite patient."

I begin to scribble notes to him.

"Yes, yes, now you calm down and let me talk. You need to rest. Your speech will come back. What is this I hear about you not wanting your children to visit you? They need to see you. They love you. It will be all right. Now, don't worry."

No, no! Don't you see? My babies will be scared. Not yet! I'm not ready for them yet! I'm supposed to be the strong one. I'm supposed to be their mother. I have to care for them and I can't right now, my thoughts raced without the use of words.

AT FIVE O'CLOCK Larry is bringing the children to see me. I have not seen them since I came to the rehabilitation hospital. I'm anxious about it. I do not want to frighten the children. Larry insisted that I see them today. I cannot speak, hear them completely, or see them to the left. I'm terrified at what that might do to my sweethearts.

They come despite my fears. I hug Betty Rae, as best I can with one arm, and we cry. Paige uses an electric curler on my hair. We begin to laugh as I write notes to them. However, I can tell that Betty Rae is upset. She lies with me in bed and I can smell the fresh grass in her hair. Mark is aloof, quiet like his dad. They try so hard to be strong like their father, but I can tell it is ripping them apart. I am glad they came, though. Not seeing me is far more dangerous to their imagi-

nations. They will be back, to see me improve. They will survive this
and so will I.

June 22, 1992   Finally, I am up walking with the assistance of a
walker. The speech therapist, occupational ther-
apist, and physical therapist are here this morning and afternoon. The
neurologist says that my comprehension is improving. I am hearing
better on the right side. I will practice by naming things in the room.

My spirits are low because Larry is working out of town. I am very
lonely. I am taken to the physical therapy department this afternoon. As long as I keep my eyes and my mind on the therapist, I
can concentrate.

I can see the other patients. I appear to be the youngest, except
for a girl maybe in her twenties; she is trying desperately to stack cones
on top of each other. After therapy the patients are wheeled into a
line. There we await an attendant for the return trip to our respective
rooms. While in line, many patients fall asleep from exhaustion. No
one speaks. Many patients have intravenous tubing and urine collection bags hanging from their wheelchairs.

Larry and the children came this evening. I smother the children
with kisses. I miss them so much that my heart aches.

Flowers and plants decorate my room. I have to write thank-you
notes. It will give me something to do.

Cora is my roommate. She is a petite, white-haired woman, probably in her seventies. She can speak, eat, and walk much better than
I. The doctors say she must go to a nursing home because she is not
a good candidate for their rehabilitation program. Both of us cry.

Rehabilitation and Setbacks

WHEN CLEO TRANSFERS to the rehabilitation facility a week after her stroke, she enters a new and different phase of her treatment. She has begun to realize how much difficulty she has performing common tasks of daily life. Contact with other patients at the rehabilitation hospital, especially during communal meals, shows her that many stroke victims are severely handicapped and helpless. Perhaps this provokes her to strike back and not become like them.

The personnel, aims, and functions at rehabilitation hospitals are quite different from those at acute care hospitals. The goal of rehabilitation is to help the person adapt to various handicaps in order to reach the best functioning. This differs from acute care, which consists of diagnosing and treating medical illnesses. Of course, some of Cleo's rehabilitation and physical therapies began during her stay at the acute care hospital while medical treatments were being given.

One of the major characteristics of rehabilitation hospitals is their use of a team approach, with many different professionals working together to help the patient recover. The first goal is an accurate assessment of what the new patient can and cannot do. Testing is often carried out by a variety of practitioners, perhaps including physicians, nurses, psychologists, and physical, speech, and occupational therapists. Speech therapists may test language and swallowing functions if abnormalities are suspected. Occupational therapists determine whether the patient can perform various common tasks of daily living and working. Neurophysiologists test cognitive functions such as memory, language, perseverance, visual-spatial capabilities, and so on. Once the testing is finished, there is usually a group meeting in which the findings are fully discussed and plans for management are agreed on. Each of the individuals then begins to work with the patient.

For a patient to overcome a deficit, it is very helpful if the person recognizes exactly what is wrong. That is his or her first step toward recovery. Thoroughly explaining the nature of the potential handicap is essential. For example, many individuals whose illness has resulted in a visual field abnormality do not recognize or understand that the problem is not in the eye and do not realize what they are not seeing. Consider a left visual field defect, one of Cleo's problems. Showing Cleo that she isn't noticing objects or words on her left and that she doesn't look toward the left is the start of retraining. A patient like Cleo is then trained to always look left and to make sure that she has looked to the farthest left edges of reading material, pictures, food, and scenes, lest she miss items on her left side. Missing things to the left presents an important problem when the patient tires and may forget to use this skill when crossing the street or just walking in a congested area.

A physical therapist may work on limb strength, exercises, and walking stability or may help train patients with paralysis in transferring from bed to chair to toilet and so on. Occupational therapists might work with a patient in a makeshift kitchen to help him or her to be able to cook again. Speech therapists might work with speaking, reading, and writing skills and may also evaluate and treat swallowing problems.

Rehabilitation hospitals also usually have specialists who can make and fit various devices such as braces, slings, and supports that can help patients move and use their limbs better. The strategies are to correct everything that can be corrected and to find alternate ways of doing things that can't be corrected. For example, a right-handed person who develops paralysis of the right hand is taught to do things more with the left hand.

Recovery usually takes much longer than the time it took to become ill, and stays in rehabilitation hospitals are usually longer than stays in acute hospitals. Unfortunately, insurers now hurry patients in and out much quicker than in the past, so that patients are often asked to do things that they are not yet ready to do. Much of the recovery is spontaneous and related to healing of the injured brain area and development of other brain regions that take over lost functions. Because many functions are represented in more than just one place in the brain, if one brain area is injured other regions can sometimes take over with time and training. This is especially apt to happen in children.

A very important component of rehabilitation is to fully explain to those in the patient's home and environment the nature of the new handicaps and how they should be handled when the patient returns home. Wives, husbands, children, other family members, and significant others can help with the therapy and should know the patient's abilities and limitations. Unfortunately, Cleo's journal shows a telltale absence of family involvement in her rehabilitation. She writes in later entries that she has been afraid of encouraging them to visit, that Larry is growing despondent, and that she feels desperate, too— all signs that the hospital may not be giving the family enough counseling, or that it isn't getting through.

DURING REHABILITATION, stroke patients often develop an intense attachment to therapy and to their therapists. Most therapists are young, active, encouraging individuals who work closely with patients. Progress often develops in these circumstances, and I've found that very often this leads to a problem: many patients begin to associate recovery with therapy and fear they will regress or fail to improve when therapy is stopped. Because therapy is aimed at specific goals, such as restoring arm and hand strength and dexterity, patients continue to vigorously pursue therapy. They may continue to practice arm and hand strengthening maneuvers for a long time after rehabilitation, for example, firmly wedding in their minds recovery from stroke and the full return of these upper-limb activities.

But the aim of rehabilitation is not to return all functions to normal. That is almost always an impossible goal. The aim is to return the patient to as close to normal living as possible. Patients without normal hand function on one side can live quite normally and do almost everything they could do before. The patient should be encouraged to broaden his or her view of recovery toward returning to the interests and routines that made up daily life before the stroke. Overemphasis on formal physical therapy in the clinic sometimes delays return to normal activities and socialization. Much of recovery involves learning to perform daily activities in one's customary locations—the home, the backyard, the grocery store. The stroke survivor must once again think of himself or herself as a person, not just as a sick care-receiver, a patient.

Heart, Arteries, and Stroke

IN ADDITION TO Cleo's dismay at the realities crowding in as she entered rehabilitation, she was there little more than twenty-four hours when she had a serious setback—another stroke. But this frightening development uncovered Cleo's specific problem—embolism caused by a clot that went through the chambers of her heart into major arteries supplying her brain.

The heart is, of course, the center of the circulatory system; it pumps blood continuously to the various vital body organs, including the brain. The heart is actually divided after birth into two rather separate circulatory systems, one supplying the lungs (the pulmonary circulation), and one supplying the rest of the body (systemic circulation). Blood returns from the veins of the body into the right side of the heart, entering its upper chamber, the right atrium. Various body organs have used the oxygen in this blood to serve as fuel for the tissues, so that the returning venous blood is blue, indicating a lack of oxygen. The blood passes through a valve between the right atrium and the lower muscular chamber, the right ventricle. The right ventricle pumps the blood through a different valve, the pulmonic valve, into the main pulmonary arteries, which supply the lungs.

In the lungs the blood is reoxygenated. When it returns to the heart through the pulmonary veins, it has a bright pink or red color, indicating its high oxygen content. The oxygenated blood goes into the left side of the heart. It enters the left upper chamber (the left atrium) and passes through the mitral valve into the left ventricle. The left ventricle then pumps it through the aortic valve into the aorta, the main artery leading to many systemic artery branches taking oxygen to the tissues. After organs extract the oxygen, the blood returns to the right side of the heart for another trip through the lungs.

The valves in the heart make sure blood goes in the right direction, and not backward from the ventricles into the atria, or from the pulmonary arteries and aorta back into the ventricles. After the chambers have contracted, expelling their blood contents, the valves close behind the contractions. The drawing on the facing page shows the heart and its various chambers and valves.

Before birth, since babies' lungs do not function and do not breathe air, a hole exists in the muscle septum that separates the left and right atria. This hole allows blood to go through the mother's circulatory system for oxygenation. At birth, this oval hole (or *foramen ovale*, a Latin phrase meaning "oval window") usually closes, but in about 30 percent of individuals, it doesn't fully close. Some heart problems are congenital, meaning that the defects are present at birth and remain. These congenital malformations may involve the heart valves. Valves can be thickened, causing narrowing of the outflow tract (called valvular stenosis), or can be leaky, allowing blood to pass in a reverse direction (called valvular insufficiency). In some patients, other holes exist between the left and right atria or ventricles. These are referred to as atrial or ventricular septal defects.

Cleo was born with an abnormal pulmonic valve that made it impossible for her right ventricle to pump blood normally to her lungs. This condition

A. *Normal heart.* Arrow indicates normal blood flow.

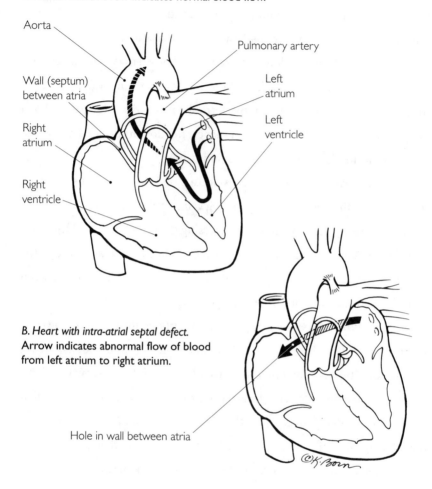

Aorta

Pulmonary artery

Wall (septum) between atria

Left atrium

Right atrium

Left ventricle

Right ventricle

B. *Heart with intra-atrial septal defect.*
Arrow indicates abnormal flow of blood
from left atrium to right atrium.

Hole in wall between atria

The normal heart and the heart with an atrial septal defect. *The heart's left and right atria (upper chambers) are separated by a wall called the intra-atrial septum. A natural opening exists within the atrial septum before birth and usually closes after birth. When it fails to close, the opening is called an atrial septal defect, and it results in blood flowing back from the right atrium to the left atrium, bringing a risk for various circulatory problems, particularly clots, which can lead to stroke. This is what happened to Cleo.*

(pulmonic stenosis) was repaired when she was 5 years old, and, in fact, Cleo had put it out of her mind completely. Apparently, as a baby, Cleo also had a hole in her heart between the right and left atria, a situation called an intra-atrial septal defect, which allowed unoxygenated blood to pass from her right chambers to the left systemic circulation, causing blue, unoxygenated blood to circulate. Hence the old descriptive term she mentions, "blue baby." When her pulmonic stenosis was repaired, I suspect her doctors decided not to risk repair of the atrial septal defect at the same time.

For many reasons, a person can develop a blood clot (thrombus) in a vein somewhere in the body, especially in a leg vein. Sitting for a long time in one position, crossing the legs and so compressing the veins, abnormal leg veins, and abnormally increased tendency for blood clotting are just some of the reasons leg vein thrombosis occurs. When the thrombus first forms, it can break loose from where it formed and go to the heart with the returning venous blood. When it goes to the right atrium, through the valve into the right ventricle, and then through the pulmonary arteries into the lungs, the resulting condition is called pulmonary embolism and can be very serious. When there is a hole between the atria, a thrombus formed in a leg vein that reaches the right atrium can pass through the hole into the left atrium and then through the mitral valve into the left ventricle. From there, the clot goes through the aortic valve into the aorta and finally into one of the aorta's systemic branches. If a clot enters one of the arteries to the brain, the embolus can cause a brain infarct. The drawing on the facing page shows the main arteries that supply the brain.

When defects cause clots to pass between the two circulations, it is called paradoxical embolism, the medical term for a stroke caused this way. Cleo's clot passed into one of the vertebral arteries in the back of her neck and then moved into the basilar artery within her head. In the brain, the basilar artery supplies the brain stem and cerebellum, and its end branches supply the thal-

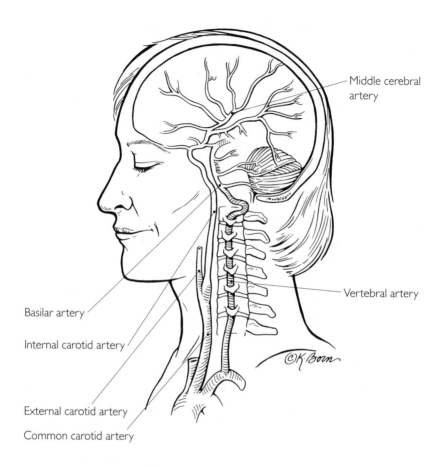

Middle cerebral artery

Vertebral artery

Basilar artery

Internal carotid artery

External carotid artery

Common carotid artery

©K. Born

The major arteries that supply blood to the brain. *The two large arteries on the sides of the neck are called carotid arteries. Each carotid artery branches into an external carotid artery, supplying the face and its structures, and an internal carotid artery, supplying the brain. The internal carotid arteries don't cross over as nerve pathways do; they take care of the brain on their own side of the head. Each carotid artery has branches that nourish the frontal, parietal, and temporal lobes, as well as the basal ganglia, internal capsule, and the eye on its own side. In the spinal column are paired vertebral arteries. They branch off from the arteries to the arms and enter the back of the brain through a large hole between the skull and the neck vertebrae meet. The left and right vertebral arteries supply the lower brain stem and the back undersurface of the cerebellum on each side and then join to form the basilar artery, a midline vessel that supplies the pons and the upper brain stem. The basilar artery divides a midbrain level into branches supplying the thalami and the temporal and occipital lobes on each side.*

ami and temporal and occipital lobes. Passage of material through the basilar artery explained Cleo's brain infarcts.

Cleo's doctors were able to determine that she had had a paradoxical embolus that damaged her brain stem, cerebellum, thalamus, and right cerebral hemisphere. This stroke affected many different strategic areas in her brain and led to her loss of different abilities.

Aphasia, Language, and the Brain

CLEO'S SECOND EMBOLISM caused ischemia in the language zone of her left cerebral hemisphere. This was an agonizing turn of events.

Language is extremely important for daily communication. The ability to use written language, to read, write, and spell, separates humans from all other species. I vividly remember the great frustration I felt when trying to obtain directions in Japan; the individuals I asked could not speak any English, and I couldn't speak, understand, or read Japanese. We simply could not communicate with each other despite earnest effort on both sides. This experience is probably a mere taste of the desperation patients with severe aphasia experience daily.

Speech mostly arises from a region surrounding the large fissure, the sylvian fissure, separating the frontal and temporal lobes on the outer surface of the brain, in the so-called dominant cerebral hemisphere. The left cerebral hemisphere is dominant in right-handed individuals and in 80 percent of left-handers. Left-handed people are more likely to have some speech functions in each hemisphere and may develop aphasia when either hemisphere is injured, but the resulting aphasia is less severe than if speech resided on only one side. Some researchers believe that women more often than men have speech representation on both sides of the brain. The drawing on the facing page shows the language areas and also indicates specialization within this region. The right cerebral hemisphere, in the same general areas, is asso-

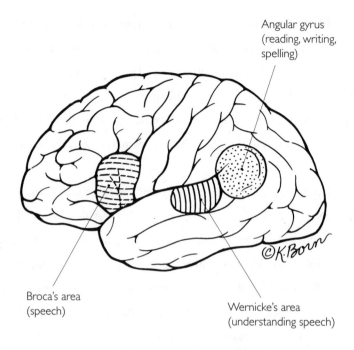

Angular gyrus
(reading, writing,
spelling)

Broca's area
(speech)

Wernicke's area
(understanding speech)

Key language areas of the brain. *In the parietal lobe, the angular gyrus region is involved with reading, writing, and spelling. Broca's area, at the foot of the lowest frontal lobe gyrus, is responsible for producing speech, and Wernicke's area, at the back of the first temporal lobe gyrus, is involved with understanding spoken language.*

ciated with adding emotional meaning to speech. The emphasis and tone of speech, and accompanying facial expressions and gestures, help send messages, and these aspects help others understand the importance to, and feelings of, the person sending the messages.

When I described how the brain works in the previous chapter, I mentioned that motor control of the face and limbs on the opposite side of the body is localized in an area called the precentral gyrus. Its lowest part is specialized for control of the muscles of the face, tongue, cheeks, and pharynx. Just below and behind this motor region is the so-called motor speech region, usually referred to as Broca's area, after Paul Broca, a French physician and anthropologist. Individuals who have damage in and around Broca's area have difficulty producing normal speech. Patients with Broca's aphasia talk less than normal, their speech takes considerable effort, and letters and words are poorly pronounced. The speech produced is usually accurate but incorrect grammatically. Writing may also be ungrammatical and telegraphic— that is, small connecting words may be left out, as is often done in a telegram. Most patients with this problem also have some paralysis of their right hand, arm, and face.

Other language problems come from damage to a region usually called Wernicke's area, after the German neurologist Carl Wernicke. Wernicke's area is specialized for the understanding of spoken language and is located toward the back of the head next to the hearing region in the dominant temporal lobe. Individuals with stroke-related damage to Wernicke's area use wrong and sometimes nonexistent words and have difficulty repeating and understanding what others say to them (Wernicke's aphasia). They may also fail to understand what they read, and they write wrong words. This is a problem of language, not intelligence. In brain regions *near* Wernicke's area, damage can cause difficulty repeating spoken language with preserved understanding of speech. Some individuals with very small lesions in the temporal lobe cannot understand words and spoken language but can hear and identify sounds and speak almost normally (pure word deafness). Others appear as if they are deaf to language and other sound input, although they jump and blink at loud noises (cortical deafness).

Strokes and other causes of damage around a region called the angular gyrus, within the lower part of the parietal lobe in the dominant cerebral hemisphere, bring about functional illiteracy. Individuals with this damage can no longer read, write, or spell correctly. Thankfully, Cleo's aphasic difficulties would recede. The embolus causing the dysfunction must have passed without leaving severe, permanent damage.

June 23, 1992 I have many visitors, and I enjoy their company. Everyone says I seem much better.

The cardiologist visits this evening and says there is a fifty-fifty chance of heart surgery. He will need to do a heart angiogram to find out exactly what is wrong. He will know by tomorrow. My spirits tumble. I am apprehensive and worried. My condition is too debilitated for major surgery. I cannot have another stroke, either.

This evening I have a slight fever and my blood pressure is elevated. The neurologist is here. He says that he is going to make me talk. He points to items and asks me to name them. I try but the words do not come out right. He points to the clock. I say, "Bad . . . boat." In my mind, I know it is not a boat, but I cannot get the word from my mind to my mouth.

June 24, 1992 I develop another rash or allergic reaction due to the seizure medication. It is early in the morning when my gentle-faced neurologist enters the room and says, "Where would you like to go and who would you like to do your heart surgery?" I cannot speak. It is the first time I am glad that I cannot. The man that I trust with my life is making this too real for me. The only decision I want to make is which day I can go home! I turn my head away from him. I am not going to accept this.

I am doing better in physical therapy. I can walk with a walker with my head up straight. Bill and his wife Sharon visit me there. As I see them, I want to stop my therapy immediately. I need to see Bill. Sarah, my therapist, says that I must wait until after the therapy session. I have to abide by her wishes. Bill and Sharon wait patiently.

Bill is 46 years old. He has a malignant brain tumor. I want to cry for both of us. At our last meeting, a lab technician was drawing blood from me and couldn't find a vein. I winced. Bill said, "I understand. They can't find a vein on me either. The chemotherapy ruins them all." I feel close to him; we have a common bond. He looks so physically fit yet speaks with a definite slowness and hesitates on each word. His hair is thin from the chemotherapy. He says he works out every day. He encourages me to fight the beast that has ravaged my body. He is my inspiration. He gives me the encouragement I need. He holds my hand and talks about his faith-life journey and our families. We end with a hug and he says, "I'll pray for you." Silently, Bill instills the message of happiness and to enjoy the little moments in life. He says to stretch my abilities not my disabilities, and always learn something new each day, pray for others, and smile more.

Bill is a former history teacher at our community high school. I remember when Sharon and Bill, and Larry and I, taught Sunday school at our church. Why? Dear Lord, why do these horrible devastating diseases strike those we love?

I cannot speak but I can hear and comprehend much better. I continue to write notes, though. I need to start talking more but the words are jumbled.

June 25, 1992 I take a shower for the first time at this hospital. It feels wonderful to have warm water, soap, and shampoo all over me. I have to use a shower chair. No one can tell the difference between the tears on my cheeks and the spray of the water. I never want to leave the safe cubicle. Showering, a simple task I always took for granted, is now a marvelous chore.

Last night I began to move my left hand and fingers more purposefully. I am obsessed with the movement and can't wait to show it off to the neurologist. My elbow leads my arm upward and my fingers can wiggle ever so slightly. I am elated over two distinct movements at the same time. If I am not straining to look directly at my left hand, it moves about in a grotesque bending action, but if I stare at it in thoughtful concentration, I can move my fingers apart and together on the bedsheets. More control, but still numb. In physical therapy I am working so hard that my leg seems to tingle from my knee to my foot. Is it possible that I am getting better already?

Both arms are bruised from daily blood draws. These vampires of the laboratory always arrive before breakfast with a bright disposition because they are not going to be pricked with the damn needle.

This afternoon God smiles on me. My speech is back! It is choppy and monotone. My head twitches when I try to speak and my eyes blink as if I am searching for the word in my mind. I have changed my vocabulary from "bad" to a more persistent and deliberate "um," as I try to get the specific word from thought to speech.

I receive a single-tipped cane in physical therapy and use it only if I feel myself falling. Sarah, my therapist, is a tall, dark-haired, thin, self-assured woman probably in her later twenties. She insists that I wear the therapy belt whenever we go walking. This wide, white canvas belt is strapped to all the patients during therapy. She says something about it being hospital policy. I am getting better, Sarah, don't you see that? I try to say.

"One step at a time," she retorts. "Remember, it takes time." Time seems to be a four-letter word in this recovery process.

June 26, 1992   The cardiologist says that he wants to do an angiogram soon. What choice do I have? I cannot avoid the issue any longer. I am getting better now. Facing this painful decision is difficult. But it is just a procedure, a test to find out the truth of my condition. I cannot afford to be scared now. The hospital staff has given me hope, encouragement, and help.

Tomorrow is Mark's birthday and I will not be able to share in his celebration. Doesn't my family need me? Larry says he has everything under control. I cannot even get Mark a card. All I can send him is my love. Does he know what he really means to me? I love him more than myself.

June 27, 1992   Today is Mark's birthday and I miss him terribly. I awkwardly attempt to use the telephone with one hand. Then a nurse writes the telephone number down so I don't forget it, and I slowly punch in the seven numbers. I try to express the words "Happy Birthday" to my son and try to talk to Larry.

Every profession has its group of undesirables, and nursing is no exception. Gloria, the nurse from hell, is assigned to me again today.

Maybe it is because I cannot complain about her. Maybe it is because I do not require much nursing care that she is assigned to me. She grabs me and props me up in bed like a rag doll. She talks to me as if I were a baby. I am so vulnerable and afraid. I do very little to provoke her, including using the telephone while she is on duty.

I remember waking up during the night to watch Gloria go through the garbage can in my room. She's reading all the notes I have written to friends, opening every crumbled ball and silently perusing each line. She is brazen enough to turn on the light during her evening read-a-thon. I feel violated. This is my personal conversation.

June 28, 1992 I feel several large lumps under my right arm and in my right groin. It must be the residual effect of a rash.

The angiogram is scheduled this morning. For the first time I accept something for pain, two pills. My body begins to react and I start to hallucinate. Is there a rat on my toes? Get that dog out of my room! While I'm being wheeled to the X-ray department for a chest X ray, it seems that a huge Doberman pinscher is stationed next to the door in the X-ray department. Could this be a medication reaction or a defense mechanism? I will rest and it will pass.

Larry, Betty Rae, and Mark visit this evening. Paige is working. I begin tape-recording visits of various doctors, with their permission, to make it easier for Larry. I cannot remember everything of importance. He listens with interest as the cardiologist speaks of the angiogram.

It seems as if I have been away from my family for a very long time. I need to be filled with every detail as I question them incessantly in my slow, monotone voice.

A nurse arrives with a long glass tube and a Q-tip. "Doctor's orders. I have to culture the rash in your right groin," she says. This is a painless and quick procedure. I want to get back to visiting with my family because their time with me is always so short.

A sleeping pill is ordered late in the evening so I will be well rested for the angiogram in the morning.

June 29, 1992   I awake to the ever-present blood tests that are required for this procedure. The intravenous tubing in my arm is disconnected and carefully closed off from the medicine bag as I dress in a fresh hospital gown. Valium is given for relaxation. I use the bathroom with help. Larry and Betty Rae are with me. It seems that I can't say to Larry what I want to say with Betty Rae present. I am not about to change the situation and frighten her.

A person wearing green surgical clothing helps me onto the stretcher and takes me down the hall, away from my family. We go past the long rows of bright lights, down in the elevator, through the double doors, and finally reach the area where the angiogram heart test is to take place.

I cannot see the doctors' and nurses' faces because of the masks, but their voices sound soothing and matter-of-fact. The Valium and the talk therapy comfort me. The cardiologist chooses the left groin as an entry site, just to be safe. This is a fine choice by me as it is my paralyzed side. Two catheters are used for this procedure. They talk me through it. They inject dye and I feel a burning sensation in my mouth, but it subsides quickly. I hear my doctor talking to assistants and nurses and I have the feeling the procedure is almost over. There is no need for worry now.

Back in my room, I have to lie flat with a sandbag weight on the entry site for a few hours. The cardiologist orders an injection for pain. My ear does not buzz and I feel calm and warm again. I say good-bye to Larry and Betty Rae and sleep comfortably.

When I awake, a nurse is placing a hot pack on my right groin. She says I have a staphylococci infection.

The cardiologist returns. He is a good-looking man. He seems to be a caring, gentle, quiet man. At times, I start laughing with him at the end of our visits. Trying desperately to tell a joke, I forget the punch line and he says it. He places his hand on mine in a "you're going to be all right" fashion and we laugh.

This time he is not smiling. I cannot read his expression. He asks me to turn on the tape recorder. I turn my right ear toward his lips to hear every word. "There is a hole between the chambers of your heart where blood clots have been going through. It's about the size of your thumbnail. This means immediate *surgery*." I turn the tape recorder off and he leaves. Later, I play the tape over and over again. Stroke, staph, surgery . . . stroke, staph, surgery . . . I can't think about this now—it's just too much.

It is therapy time. I just cannot do this today. I am brought back to my room. Somehow they understand.

So Many Tests

CLEO'S JOURNAL speaks of many tests. An important one she mentions in connection with both her strokes is magnetic resonance imaging, or MRI, one of the most valuable diagnostic tests that doctors use. The last twenty years have brought dramatic improvement in the technology available to diagnose stroke patients. Using computed tomography (CT) and MRI, doctors can now safely and quickly determine what type of stroke the patient has had, ischemic or hemorrhagic. They can see where and how much the brain is damaged and the presence, nature, and severity of any abnormalities of the blood vessels that supply the brain. Electrocardiograms, echocardiograms, and heart rhythm monitoring can also show whether there are heart abnormalities, and blood tests can show whether blood abnormalities caused or contributed to the stroke. Knowing exactly what is wrong with the stroke patient allows a physician to select the best treatment.

Probably the most important clues to where a stroke has happened in the brain and what caused it come from the history the doctors take from the patient and their physical and neurological examinations of the patient. Many patients think that these examinations have been replaced by the new technologies, but nothing is further from the truth. Examining a patient is the only way to know what he or she can and cannot do. The doctor should always begin by taking a history of how the stroke came on and developed later, previous events and illnesses, various risk factors, and what symptoms and problems the patient noticed. Examination of the pulse, heart, and blood vessels of the arms, neck, and legs gives the doctor clues to abnormalities in the heart and blood vessels. During the neurological examination, the doctor checks speech, memory, reading, writing, thinking, walking, vision, hearing, strength, coordination and feeling in the limbs, and various reflex functions. The information from the patient's examination and history helps the doctor plan the appropriate tests.

The tests devised during the last two decades make it possible for doctors today to obtain detailed images of the brain. For stroke, we use two general types of brain imaging tests—CT and MRI. CT uses ordinary X rays and computers to create thin sliced images through various levels of the brain. MRI uses magnetic energy to generate images of the brain. Both tests are safe and do not hurt. Each is done with the patient's head positioned in a machine; with MRI much of the body is also enclosed. Patients must remain still if the machines are to generate quality diagnostic images. The MRI machine does make rapping noises—Cleo refers to this—and people who are claustrophobic have difficulty staying motionless in it. To gain more detail about the brain image, doctors sometimes order an intravenous injection of some contrast-producing material. With CT, this is usually a dye that contains iodine; with MRI, it is a chemical called gadolinium. Some patients can have an allergic response to these materials, especially to the dye used for CT contrast.

CT and MRI allow doctors to distinguish brain hemorrhages from infarctions. Look back at the CT images on pages 26. On CT scans, hemorrhages appear white and infarcts gray or black, making it quite easy to distinguish between the two main stroke categories. MRI produces more different brain section images and shows the brain stem and cerebellum better than CT. The brain images show not only whether the lesion is a hemorrhage or infarct but also exactly where the injury is, how large and extensive it is, and whether there is brain swelling and pressure buildup caused by the infarct or hemorrhage.

In some patients with brain ischemia, CT and MRI scans can be normal, indicating that the brain has not been irreversibly damaged—that is, not yet infarcted. In these patients, doctors find neurological signs when they conduct an examination, but no abnormality appears on the brain images to explain the dysfunction. Remember Cleo's CT scan on the day of her first stroke? It was negative, meaning that no abnormalities were found, despite the fact that she had numerous neurological dysfunctions. Many people think that imaging scans are infallible, but they certainly are not. Patients with normal scans will return to normal function if the ischemia is reversible; in many patients, an infarct will develop later, and then CT or MRI scans will show the area of infarction. This is the reason doctors often order scans later, especially if the initial scan was normal. Knowing where the problem is in the brain allows the doctor to know what blood vessels supply the affected region. These vessels can then be examined.

Having identified the brain problem, doctors naturally turn to testing the arteries in which abnormalities could have caused the stroke. Pictures of arteries and veins are referred to as *angiograms* and the process of taking the images as *angiography*. CT angiography (CTA), sometimes called spiral CTA, magnetic resonance angiography (MRA), and ultrasound are commonly used now to analyze the nature, location, and severity of any changes in the blood vessels supplying the brain. These tests are all quite safe and can be performed quickly either in the hospital or out. When the infarct is in the thalamus, cerebellum, and temporal and occipital lobes, as it was in Cleo, the doctor knows that the problem must lie along the path to that region, which includes the heart, the aorta, and the subclavian, vertebral, and basilar arteries. Ultimately, material must have blocked the end of her basilar artery and extended into the right terminal branch of that artery, called the right posterior cerebral artery. But if the problem were within the right middle cerebral artery, then the path to be analyzed would be the heart, the aorta, the right carotid artery, and the right middle cerebral artery.

MRA can be performed at the same time as MRI. MRA produces an image of the arteries and veins, which is possible because flowing blood can create an image. When the flow of blood in an artery is blocked, then no image of the vessel appears. Another technique, CTA, involves the injection of dye, which allows an image of the arteries to be produced; it can be performed with a CT machine that rotates. CTA can be done at the same time as CT.

Ultrasound, sometimes called Doppler after Christian Doppler, who discovered the principle of ultrasound while studying astronomy, is another very effective way to safely study blood flow in the arteries. In an ultrasound examination of the carotid and vertebral arteries in the neck, a technician or physician places a small probe over areas in the neck. The most familiar machines, called duplex ultrasound, produce an image of the artery as well as graphs of the speeds of blood flow in the artery being imaged. These two types of ultrasound information allow doctors to tell whether an artery is normal, blocked, or narrowed. If the artery is narrowed, the degree of narrowing can be estimated rather well. Some laboratories use transcranial Doppler (TCD) ultrasound. In this technique small probes are placed over the eyes, the back of the neck, and the temples, places where there are natural holes in the bone or where the skull bone is thin. Blood-flow velocities in the various arteries within the skull are also checked for narrowing, occlusion, or increased or decreased blood flow. When an artery in the neck is narrowed or blocked, TCD of the main branches of that artery in the head can reflect the impact of the neck disease on blood flow to the threatened region of the brain.

In some hospitals, radionuclear scans called SPECT (for single-photon-emission computed tomography) are used to estimate blood flow to a region. After a chemical tagged with a radioactive substance is injected into a vein, the brain is scanned with a special machine that detects where the injected substance has gone. This helps reveal relative blood flow to various parts of the brain, but unlike MRA, CTA, and ultrasound, it does not give images or direct information about the condition of the supplying arteries.

When these noninvasive tests do not give sufficient information about the diseased arteries, physicians may order a cerebral angiogram to be performed by a radiology specialist. This test is more invasive and carries a small but definite risk of complications. When it is ordered, the information needed is important enough to warrant the risk. The radiologist places a catheter in one of the arteries of the thighs or arms and threads the catheter into the neck arteries to be studied. The procedure is performed under visual control by means of a TV-monitoring fluoroscopic screen. Dye is then injected and a series of rapid-fire X rays are taken, producing pictures of the dye as it passes through the arteries and veins. Pictures of the inside of the arteries are made that can show narrowing, obstructions, aneurysms, and vascular malformations. Complications most often relate to an allergic response to the dye, inadvertent injury to the artery into which the catheter is placed, or the dislodging of a plaque or clot from an artery so that it passes into the brain. In the hands of the best-trained and experienced neuroradiologists, these complications occur infrequently, in one or two patients in a hundred, and the great majority of complications are minor and temporary. Still, doctors try to avoid angiography if the preliminary tests give enough information. The more the doctor knows about the patient's abnormalities, the more precisely he or she can choose treatment.

The heart is an important actor in all functions of the circulatory system. Ultrasound of the heart (called echocardiography) can yield pictures of the various heart chambers and their functioning. The echocardiogram is performed by a technician and a cardiologist with an ultrasound probe on the chest or by having the patient swallow a string-like device containing an ultrasound probe into the esophagus. Much of the heart is better seen from the back, through the esophagus, than from the front. Sometimes saline bubbles are injected into an arm vein to see their passage through the heart. In Cleo's case, bubbles that were injected could be seen to pass from her right atrium into the left atrium through a hole in the wall between the two. Electrocardiograms and monitoring of heart rhythms are other important heart tests often used in stroke patients. Heart rhythm abnormalities can lead to clots forming in the heart and embolizing to the brain.

Lastly, blood tests are important. The blood contains very small cells called blood platelets that play an important role in bleeding and clotting. The serum of the blood contains various protein coagulation factors that also relate to blood clotting. Abnormalities of either the platelets or the coagulation factors can potentially lead to a heightened tendency to bleed or clot, which can cause or facilitate either brain hemorrhage or brain ischemia. Blood tests measure the number of red blood cells, white blood cells, and platelets. Other tests tell if the blood clots normally, and determine the amounts of various proteins in the serum that affect blood density and clotting. These tests are especially important if an increased bleeding or clotting tendency is suspected and if medicines are going to be given that affect blood clotting.

June 30, 1992 A shower would feel so good right now. I never felt so helpless. No shower today, just a bed bath. I am so tired of bathing in my bed just because the occupational therapist wants to see if I can wash my right armpit with my left hand! Put me in the shower, please. Give me a long-handled brush and I will figure it out! The occupational therapist wants to see if I can dress

myself. If I do, then will she undress me and give me a shower? I think
we have reached a point of impasse.

My speech therapist enlarges the print on a few pages of a book I
want to try to read. I read aloud to practice my voice inflection and
power. In therapy today I recite one of my favorite recipes from memory, slowly searching for the correct word and sentence usage. I seem
to be healing.

On rounds, the cardiologist tells me the surgery is scheduled for
the day after tomorrow. Claire is visiting when I am told. She telephones Larry to come immediately. She must have seen the concern on
my face and decided I should not be alone. Larry visits late in the
evening and listens to the tape. My eyes begin to tear and he knows I
do not want the surgery. Sternly he says, "What do you want? Do you
want to end up in a nursing home, not knowing what the hell is going
on, or worse yet, dead? Do you want to be the mother to our children
again? Aren't you glad they have the technology today to repair the
damage? A few years ago, who knows? Maybe they would have said,
'Sorry, we found the problem but we can't do a damn thing about it.' "

Maybe he is being optimistic. Maybe he is being an insensitive ass.
No matter, he is right. What choice do I have?

July 1, 1992 The neurologist assigned to my case at the rehabilitation hospital, a stocky, dark-haired man, tells
me in a matter-of-fact way that he is not at all concerned about the
anesthesia affecting the stroke. He explains that I have the best team
of surgeons in the metropolitan area. Well, it is little consolation. He
is not having the surgery! I know I am a high-risk patient.

I am given a shower today. No fighting or begging. Have the nurses read my mind? This would definitely have been my last request, if I were to make one. The only difficulty we encounter is getting me out of the shower and dressed in time for therapy. In physical therapy I walk by myself with no cane and no limp. I concentrate on every movement. I am so proud of myself.

A thoracic surgeon visits me this afternoon and explains the surgery more specifically. He takes one quick look at my right groin and immediately cancels the surgery for tomorrow. "We can't do surgery with a staph infection! We have enough negatives going into this operation without adding a serious infection to them."

The internist and cardiologist begin rigorous routines of antibiotics, both orally and intravenously. I have intravenous tubing in both arms. Each arm is covered with bruises from blood draws and intravenous tubes. The nurses continue to hot-pack the infection.

It is difficult to keep my spirits up now. I feel as though I am unable to move. I wish that I could see all the doctors at the same time. I would say, I'm having enough problems with memory without controversy on your part. Tell me this surgery isn't risky. Tell me it's a walk in the park. Tell me what I want to hear. I feel vulnerable, exposed, and unable to control my environment.

July 2, 1992 I continue to battle the staph infection while attending therapy sessions. As I exercise with Sarah today, we speak of feeling stronger, and I let a part of my former self come through. "Unfortunately, you must check your vanity at the door. You need to focus entirely on getting better. Not makeup, not hair, not beauty, just working to get better," Sarah says. I agree, yet I know I must be improving if I can even dare think of my appearance.

Sometimes I notice a therapist hoisting a gown or pajama pant leg up to thigh height to observe a knee joint or see if a leg has atrophied. I sit helplessly by and watch elderly men who have had strokes, young people recovering from diving accidents, children who are the victims of violent crimes, elderly women who have fallen, and teens injured in car accidents. I see tears of frustration and tears of accomplishment. I know that I must keep my mind on little improvements and the work before me. I must never let go of this thing called hope.

Larry and I need each other and the sincere conversation we are trying to have this evening. It is about fears, real and unreal. He sits beside me in bed and holds my hand as we talk about each other, the children, our finances, and the surgery. We are under so much stress. He seems very optimistic about the surgery. I'm afraid. I believe he may be too, but he refuses to admit this. He has to be strong, doesn't he? He did admit that our finances are in a shambles. He is doing what he can to keep things from going from bad to worse. We need to sell the house, but the market is bad and he projects that we will lose a great deal on the sale. "Please stop crying," he begs. "I can't leave you like this."

"Then don't."

"I have to, Honey. I have things to do at home."

July 3, 1992   Diane is a roommate of mine. She is a dark-eyed, talkative 23-year-old with long, silky black hair. She had a fainting spell at home and her mother brought her to the hospital. Diane has two children, a 2-year-old and a 6-month-old. She was admitted for observation and a few tests. She shows me pictures of her babies. She talks incessantly. I understand that she needs someone to listen to her, and I am certainly good at that.

In our small, cramped hospital room it is impossible not to overhear Diane's doctor's conversation. Diane has a brain tumor. Surgery is needed to tell if it is malignant or benign.

Diane didn't say a word after that. The next day they do the surgery. It is malignant. I weep silently.

The Effect of Other Patients

DOCTORS, PATIENTS, AND CAREGIVERS often underestimate the positive and negative impact of other patients on a patient's attitudes and recovery. Cleo has already mentioned her elderly roommate, Cora, and her friend Bill, who is fighting a brain tumor, and Diane. Clearly these friends who are also ill are important to her.

The psychological effects of other patients are complex. Most simply send the message "You are not alone." Many also offer advice and support. After all, they too have had their own medical problems, and they know what it is like. Patients who have successfully recovered from a stroke can be role models, offering evidence that it is possible.

What about contacts with patients who have more severe illness and those who have not done well? One might expect these patients to be "downers" for those who are less sick. In practice they usually are not. Patients with stroke and other major illnesses have two very common reactions: they often feel sorry for themselves and they are often angry that this undeserved calamity has fallen upon them. What did they do to deserve this fate? Seeing that others, probably equally undeserving of bad fates, have also been stricken, and that others may have even more desperate health problems, helps them put their own illness and condition in perspective. It also takes their attention away from themselves and directs it toward others. When I care for patients who dwell on why they have been "chosen" by an uncaring God to be struck down in the prime of life, I often suggest that they read the book *When Bad Things Happen to Good People*, by Harold Kushner. This book discusses the problem of blame for illness, an issue as old as the Book of Job in the Bible.

The mere fact of becoming ill and having a potentially serious illness such as stroke, changes a person forever. To most healthy individuals, illness is something other people get. Our daily lives are complex and demanding enough that the idea "What if I become seriously ill?" is so daunting that most of us exclude the possibility from our minds. The other extreme is to focus too much on minor symptoms. But if and when sickness comes, it is impossible to keep the threat of more future illness totally out of our consideration. Patients view themselves differently than they did before they became ill. They also view other patients differently from before. They feel more personal attachment, more compassion, more understanding. I know this from my own personal encounters with serious illness. Having been a patient with a potentially serious illness, I believe, made me a much better doctor.

July 4, 1992 The holiday, and my family is continuing our tradition of spending the day at our friends' cottage on a beautiful lake where they will picnic and the children can swim.

No shower or bath for me today. Most of the staff is on holiday. I try to joke with staff members because I am sure it is not easy to give up their holiday. Two nurses move my bed close to the window. We watch as the city's fireworks light up the evening sky.

Diane is in intensive care.

July 5–7, 1992 The intravenous line in my left arm dislodges from the vein during the night, and my arm appears the size of a small melon. I cannot feel this happening. The intravenous team tries seven or eight times to start a new line, but they fail. I am told that there is so much scar tissue it is almost impossible to find a vein. A surgeon states that if we lose the intravenous line in my other arm, the medical team will have to resort to other means to get the much needed antibiotics and blood thinner administered. What other means?

The staph infection is not clearing up as fast as it should, and the doctors are getting frustrated. They call in an infectious-diseases doctor.

I am moved to a different room. My new roommate is an eighteen-year-old high school senior who was in a car accident. Julie is in a coma. Her mother stays with her constantly. She comforts, soothes, and loves her beautiful sleeping daughter. A drunk driver hit Julie.

I am angry at the injustice of it all. I am depressed because no one can make her better except God.

July 8, 1992 Tomorrow is the big day. The cardiologist says that it is best to repair the defect immediately. The risk is too great to wait any longer. I will leave the neurological floor at six in the morning and have surgery at eight-thirty. It seems as though every doctor in the hospital is in to see me today. No therapy because there is so much preparation; blood tests for clotting time, arterial blood tests, and of course the medical staff will not forget the laxatives.

My neurologist says that I am ready for the surgery. I pray he is right! The anesthesiologist says he will be with me at all times, monitoring everything. The infectious-diseases doctor will watch and advise.

I cannot spend today staring at the walls. I must write to Larry.

To my beloved husband,

What do I say to the man I have spent half my life with? The words "I love you" are too inadequate. We have grown together, experienced life together, comforted each other in the death of loved ones together. We have given strength to each other even when we felt we had none for ourselves.

If love means caring, dedication, being quiet when you need to be, laughing at jokes you heard a thousand times before, then I know I love you. If love means supporting someone through an illness, staying with someone when you feel like leaving, not arguing with someone even when they want to, loving someone so tenderly and admiringly, then I know you love me.

I have known your humanness. I have known your sensual touch. I have known your soul and I thank you for allowing me to be that close to you. You have always been a great father and provider. The children adore you. You are strong, yet guide the children with a tender hand. You have always done what needs to be done. Now the children need you and you need them.

In time, I hope that you will again meet someone. Know that it's all right to continue living.

I love you,

Cleo

I carefully slip the note into my bedside table. Am I losing control of my life? I am now in the hands of God and the skilled surgeons.

CHAPTER 4

Heart Surgery and Wrestling with Rehabilitation Again

July 9, 1992   I am given two injections. Larry, Paige, and Mark are with me. Larry tells me that he thought it best that Betty Rae spend a couple of days with her friend and the friend's family. Mark finally breaks the silent vigil with a "See you later, Mom." In his voice is every assurance in the world that everything will be all right.

In the preoperative area I see my surgeon's eyes over a green surgical mask, and I know the exact procedure this surgery will take. The years that I had worked as a nurse and assisted in surgery rushed through my mind. No injection can disguise the fear. The success rate in this type of heart surgery is wonderful, but I also know I am now considered part of their job. I am losing my identity. As he pulls the tourniquet tightly around my arm, I mumble something about my instinctive desire to go home and to forget about this; anything to divorce myself from the situation. He has someone else, probably from the anesthesia department, working on my left side. The surgeon gives this person a nod and says to me, "Relax. You know too much. You're . . . go-ing . . . to . . . be . . ."

I WAKE IN THE INTENSIVE CARE UNIT on a respirator. I know I have a tight oxygen mist mask on my face and a gastric tube in my nose. I can hear the sound of the rhythmic pressure. I try to make a noise. "As soon as you're breathing better on your own, Honey. In about eight hours we'll remove the respirator."

With my right hand I begin to paw my body, two chest tubes and a urine catheter. "You can't pull them out!" The nurse administers something in the tubing. "Now, that should help you feel more comfortable," she says as I drift back to sleep.

I awake again, this time more agitated about the respirator. A nurse suctions secretions from my throat. "Four more hours dear, just relax."

The time has finally arrived to get me off the respirator. The nurse places some pink liquid into the mouthpiece of the respirator. I cough and she suctions and pulls the tube out.

I remember the preoperative instructions about how to use the deep breathing machine and how to hold a pillow tightly over the incision for coughing. These tasks seem impossible for me now. I am in and out of sleep.

Every four hours, the inhalation therapist begins mist treatments to prevent pneumonia. They encourage me to suck into a breathing apparatus to keep my lungs clear. I am not good at performing the task.

It must be late at night because it is quieter in the unit. I begin coughing spastically. It is hard to hold my pillow tightly due to my left arm and all the tubes. I cannot sit up. I cannot move. Where are the nurses? Can't they hear me? Help me, somebody, I'm going to choke to death! Hurry! Suddenly a nurse turns me on my side and begins to suction the secretions. I feel a burning pain through my left arm as if it is in a tourniquet. It is the blood pressure machine. For the first time I feel something from my left arm. I know that I am going to be okay.

I spend three days in the intensive care unit. I keep thinking of the other patients in the unit, the brother and sister in the car accident who are now quadriplegics.

I can feel something in my left arm. I know that I am one of the lucky ones. I am becoming more aware of my surroundings as time passes.

A blond nurse who appears to have practically no eyebrows or eyelashes whips the curtain back. "Well, let's get your tubes out and get you ready. You're going to be transferred to the cardiac floor. The cardiologist will change the tubing in your neck to an intravenous type. That way you won't have to be stuck with needles so often." She washes me, dresses me in a clean gown, and places long white elastic stockings on my legs. She says they are to prevent clotting.

Finally the cardiologist arrives. The intravenous tubing in my neck is changed and he stitches a tube called a triple lumen in place. "Now, don't be trying to talk while I'm doing this. It will only take a minute. You're doing just fine. We just sewed up the defect in your heart, no valve replacement, just a simple patch job. There now, don't you feel better? There may be a little setback with the stroke symptoms but you should regain what you had before surgery," he says so convincingly.

I am transferred to Room 633, close to the nurses' station, and the nurses keep me comfortable with morphine.

July 14, 1992   An unfamiliar thoracic surgeon visits and states that my surgeon is taking a well-deserved day off and that he is an associate. He has come to remove the chest tubes. He clips the sutures that hold the two half-inch tubes below my midline incision. I am halfheartedly amused at the way he places one hand firmly around the tubes and the other on the incision and looks me squarely in the eyes. "This will be over before you know it. When I count to three I will pull them, so get ready!" I think, "Such drama! Just get on with it." He pulls them, fast and hard. It feels as though a football player just tackled me and knocked the wind out of me. It seems like all my guts were just pulled out on those tubes. He dresses the wound as I lie breathing painfully. I memorize every feature of his face so that I will never again take his words so lightly.

I spend the day drifting in and out of sleep. I dislike being awakened every four hours for the inhalation therapist's breathing treatments. I am told I usually fall asleep during them. I cannot lift my head up. There is a sharp pain on my right side, and I breathe very shallowly. I have an oxygen cannula in my nose. The constant blood tests are easily obtained through the triple lumen in my neck, and I am not disturbed by the procedure.

"We're going to have to give her three units of packed cells stat, then check her lungs down in X ray for possible pneumonia." I can hear the voice close by me. Who is it? Can it be the cardiologist? My body is too heavy to move. Let me sleep.

I am taken to the X-ray department where a huge, round, cold machine awaits my arrival. The X-ray personnel inject dye into the tubing in my neck and turn me around to photograph every aspect of my lungs. At one point during this procedure I notice my original cardiologist putting another patient through her paces on the treadmill. I begin to sob as I remember when I went through the same test for him. Why couldn't he have seen the defect then?

July 15, 1992 I am worried about Betty Rae. I have not seen her since I got out of the intensive care unit. Larry tells me that on Sunday, at her softball game, she was hit by a ball and got the wind knocked out of her. Two nurses, mothers of her teammates, called an ambulance. She went to the community hospital. Larry said that she was okay. She just got a bad bump. If this is true, why doesn't she come to visit me? I need to see her now. I have to know that she is all right!

My sister Carrie and Aunt Jacqueline visit me this afternoon. I know that they have been busy caring for Brian. I need to know how Brian is doing, as his disease is terminal. Carrie is four years my junior. She has long dark hair and an olive complexion.

"How is he?" I ask.

Just then Larry, Paige, and Mark come through the door.

"Everything's fine," Carrie whispers. Carrie hands me a sculpture of a loon to remind me of our cabin on the lake. After everyone leaves, Larry tells me that Brian died around the day of my heart surgery. He says that he had asked Carrie not to tell me until I was stronger. Am I stronger now? I don't feel strong. I feel as if the surgeon is back. Instead of removing tubes, my soul is ripped out.

I missed his funeral service. Maybe our souls touched. His spirit comforts me. I will recover faster with his help. He never knew that I had a stroke. I ask the nurses for the hospital chaplain and we pray together. God grant him eternal peace.

July 16, 1992 I ask the staff to put a sign on my door to "Please keep door shut." There is no reason why I must hear elderly patients calling "Lorraine, take me home" over and over again all night long. I want to comfort them but I cannot, because I am grieving, too.

Sleep does not come. Maybe it is the energy the blood transfusion has given me. Maybe it is Brian and Betty Rae that I cannot get out of my mind. It is well after midnight and I am wide awake. A mild sedative is ordered. I decide martyrdom is not on my agenda for tonight. I seem to have just fallen asleep when someone begins nudging me and shining a flashlight in my eyes, saying, "Were you able to see to your left before the stroke?"

I hesitate for a moment to orient myself to the surroundings, thinking it may be morning. "Yes," I answer.

The light speaks again, "Was your left side numb before the stroke or did that just happen tonight?"

"The stroke," I reply.

She continues to question, "Could you hear to your left before the stroke?"

"Yes."

The day shift had been kind enough to place a large sign on my door that very afternoon. It reads in big red print "Please Keep Door Shut." As the nurse with the flashlight departs, she leaves the door open. "Shut the door!" I scream within my head, unable to respond quickly enough before she leaves the room. Ah, damn! I'm awake the rest of the night, with my eyes fixated on the bright white hall light, while listening to the moans of other patients.

Early the next morning, a staff nurse I am not familiar with enters the room. She tells me that she is assigned to me for the day but she has two other patients. She removes my urine catheter and places it in the garbage can. "I'm going on break or I'll never get one. I'll work with you later." She never comes back.

A male nurse's aide washes me and helps with my food trays for the day. The cleaning person empties the garbage. Out goes the catheter bag without being measured or charted. After that, the doctor feels I must be retaining too much liquid because he orders a diuretic. I use the commode stationed by my bedside. I am starting to become oriented to my surroundings. I do not like having to depend on others.

Many times I think of the Scarecrow, Tin Man, and Lion in *The Wizard of Oz*. I need my brain to mend itself. I need my heart to heal. And I certainly need courage to fight. I remember what the Wizard said as he handed the Tin Man a ticking watch for a heart: "Remember, a heart is not judged by how much you love, but how much you are loved by others." I wonder if I am still loved by others.

I miss my family and friends. I know their prayers and thoughts are with me through my recovery by the many cards, flowers, and personal visits.

July 17, 1992 It is early morning and every once in a while I feel a sharp pain in my lower right side. I must still have some pneumonia. Respiratory therapy continues the mist-breathing treatments four times daily. I hope that all this will help soon.

Last night I had a very vivid dream. Two young girls were playmates. However, both were physically and mentally challenged. They played tennis together and enjoyed each other's company. A young boy came to the tennis courts one day and said to one of the girls, "Don't you know she's letting you win? She is purposely playing slow because you can't keep up!" The little girl, who had heard the comment, asked her girlfriend to teach her the proper way to play tennis. Reluctantly, her friend obliged. After many hours of practice the little girl finally learned the rules and how to play tennis the correct way. The girls became even closer friends.

The girls spent a day together at the home of one of the girls, a western ranch. The little girl who had learned to play tennis spent time in the barn petting a mother cat and her kittens. As she looked up, the boy was there. "Don't you know they drown kittens on farms?" he said. She hid the cat and her kittens under the straw in the far corner of the barn.

Soon the three children went on a trail ride on three gentle horses. It was getting late and the father began to worry about the children. He was about to send a rider after them when he saw them coming down the trail, the boy in the lead, then both girls riding together. They were leading one of the horses because it had a hurt leg.

The father shouted, "Shoot the horse. It's no good to anyone now!" The ranch hands agreed that it was the only thing to do. As the horrified children held the reins of the horse, the father shot it. The little girl ran to be near her kittens, but only one remained. She held the kitten and rocked back and forth and thought of a different world, a world full of love, gentleness, and laughter. She wanted to go home.

THE HOSPITAL WORLD is a sterile, unfeeling world. I want to feel the soft fur of a kitten and hear its purr against my breast. I dream of going home.

I SUPPOSE THAT EVERYONE is handicapped in one form or another. Some people are physically or mentally challenged. While others are ethnically, racially, culturally, or socially challenged. I will practice looking inside myself to find the young child that needs healing.

IT IS NOW THE EIGHTH postoperative day, and I want my cardiologist to adhere to the bargain made with me about being discharged from the heart floor after eight days. They usually send patients home within a few days after surgery.

The doctors order a pulmonary function test. The technician explains that by blowing and sucking on a mechanical tube during this test I am using only 40 percent of my lung capacity. After the inhalation treatments, it indicates a significant improvement in my breathing ability. The pneumonia is under control. I need to get out of bed more and try to breathe deeply. I will need every bit of strength. The cardiologist tells me that in three days I will be transferred to the rehabilitation floor. This will be my last step before going home.

Dreams

DREAMS HAVE FASCINATED scientists, scholars, and all dreamers since time immemorial. Who cannot recall being intrigued and impressed about the facility that Joseph showed in the Bible in interpreting the dreams of the pharaoh? Psychiatrists have based many theories on dreams and their contents.

Clearly, in order to have a dream, the idea or content of the dream must be somewhere within the mind of the dreamer. Cleo's dream is filled with threatened, fragile beings—mentally challenged children, pets, a lame horse, all in imminent danger of being killed and injured. Cleo was clearly feeling fragile, damaged, and vulnerable. Her feelings of decreased self-image and her fears emerge in the dream.

Many biological activities contribute to perception and conscious aware-ness in the nervous system. These activities are performed mainly at the level of the cerebral cortex, where thoughts are processed. Other nervous system tasks are automatic (such as breathing, heartbeat, and the control of blood pres-sure) or reflex (such as tendon reflexes, withdrawal of a limb from a pinch, and flaring of the skin when scratched by a sharp object). These are performed at lower levels of the nervous system, such as the spinal cord and the brain stem, and take place without conscious awareness. They continue even when the cerebral hemispheres are severely damaged. The control of sleep and dreaming is probably centered in special nuclear circuits in the upper brain stem, but the cerebral cortex in some regions is probably also involved. Dreams can occur at night or during the day. Individuals usually know when they are daydreaming and can get back to reality when stimulated and when they desire. Drugs and sleeplessness can make it difficult for some patients to know if something actually happened or was part of a dream. Some patients with lesions of the upper brain stem involving the brain region called the reticular activating system (shown in the drawing on page 37) report events, hallucinations, and dreams as if they were actually happening and were real. The intricate involvement of dreams with sleep, and the separation of experiences at a somewhat subconscious level from conscious willed behavior and actual perceptions, have been a fascinating challenge to unlock. Some aspects of the anatomy and physiology of dreaming remain unsolved mysteries.

Cleo's dreams of animals being injured and mentally and physically challenged individuals being stressed shows her feelings of fragility. She clearly identifies with the animals and the girls.

July 18, 1992 Today is an anniversary of sorts. I have spent one month in this hospital. I have lost thirty-eight pounds since this ordeal began. Rumors are flying that the nutritionist has ordered tube feedings if I continue losing weight. I am not doing it on purpose. I am not hungry. Maybe I ate for all the wrong reasons. I have never had a hunger pang in my life. I ate to socialize and dined at many restaurants. As a family, we would stop at various fast-food places. We enjoyed eating pies, ice cream, or some kind of dessert. The hospital is a controlled environment, and it is teaching me healthy eating habits. I can't wait to get home and prepare a family meal that will be healthy, colorful, and appetizing.

The telephone rings. Larry is coming with Paige, Mark, and Betty Rae. I talk with each of them slowly, pausing between phrases so as not to sound too anxious. I am working in speech therapy on my voice inflection, and I try to keep my voice down to an even tone. "Hey, Mom, would you like me to bring my hair curler and fix your hair?" Paige asks. "Maybe we could help you do your makeup too!"

"I would love it, Paige," I say, almost crying. I cancel my morning therapy in order to give myself time to clean up before they arrive.

A nurse surprises me by allowing me to take a bath. However, it takes two nurses to get me out of the tub because I can't use my left side yet.

I lie in bed wearing a crisp, clean gown, waiting for their arrival. When they arrive I hug Betty Rae as if I had not seen her in months. "Come on, Mom, let's get you beautified," says Paige. Larry and Paige wheel me to the small porcelain sink in my room and begin to wash my hair. My forehead rests on a towel on the sink as they use a water pitcher to rinse the shampoo. Mark lies peacefully watching television in the empty bed in the room. They brought clothes for my transfer to the rehabilitation unit.

I love seeing everyone together. It has been such a long time since we have been a family.

This evening the doctor orders my intravenous fluids discontinued. The nurse puts injections of heparin in each neckpiece, disconnects the long tubing, and leaves the sutured tube in my neck. I am free from tubing, dripping liquids, and intravenous stands.

July 19, 1992 I have a rough night and morning. Icy cold pains electrify the left side of my face and my left shoulder and arm. "Someone please give me a stronger pain medication! This acetaminophen isn't working," I beg the staff. I think that I overdid it yesterday when I tried to walk in the halls with my family. Healing takes time.

Larry is here over his lunch hour, and again we begin to share some tough issues. He is taking one day at a time, but it is getting very difficult. Finances, single parenting, Paige's college, hospital bills, everything bothers him. I want to help him but I feel so useless. I think of what our life together will be like once I am discharged. We cry and laugh together. We desperately need each other.

I am transferred to the rehabilitation unit today. The intravenous team has orders to remove the triple lumen in my neck. This is the last bastion of my heart surgery. Although they may have to draw blood every day to regulate my blood thinner, it will not be forever. As I dress myself, I know that although it is impossible to hurry the healing process of the stroke, I am finally returning to the rehabilitation floor. After more therapy, I have nowhere to go but home. I am ready to work. This time I'll make it. Thank you, Lord.

"Your coach awaits, Madam," says a friendly voice jokingly. There in front of me is the man who weighed me each day, the man who gently washed my arms and face, the one who assisted me to the bathroom when I needed help and would just stop in my room to say hello when I was lonely. My chariot driver is the male nurse who originally met me when I came into this hospital. My opinion has changed. He has become a dear friend. The words "thank you" seem so inadequate. I hug him.

Larry came this evening and we went to the dining room for supper with the other patients. Only three elderly men are there. All three patients seem to have had strokes. Only one man looks like he needs assistance eating. It is all right for me this time, but a real revelation for Larry. He does not say a word but attempts to be cordial by his facial expressions. After supper he takes me for a ride around the unit in the wheelchair. He seems distant, quiet, and reserved. "Good-bye, Honey." With a quick peck on the cheek he is gone.

July 21, 1992 For the second day I dress myself. Today I am able to put on the long white elastic TED (thromboembolic disease) stockings by myself.

In physical therapy, Sarah takes me outside to practice getting in and out of a car. We go for a short walk up and down ramps. We sit outside and talk. It is a perfect summer day with white puffy clouds in the sky. I see children in the playground across the street and hear their squeaky laughter. A few robins warble their summer song while a cool breeze blows in my face. I can smell the earth, grass, and flowers of the season. I enjoy every moment of it.

I think about the child within me today. I remember when I first learned to tie my shoes, button my buttons, and dress myself. It was no small accomplishment. I must relearn these basic tasks again. I think of the children that were in my care for so many years and how I taught them to put on their jackets, socks, and shoes. Now I find in myself the same awkwardness in coordination and slowness in physical motions as the children. Sometime I may again be the teacher, but for now I must satisfy myself with being the student.

July 22, 1992   I hurt last night. I hurt this morning. The doctors don't want to give me any strong pain pills because it affects the brain. I wash up, brush my teeth, dress, but I hurt. I will tough it out. Sometimes I feel better and make the most of it. No day is ever uneventful or all bad. I walk farther than ever before. I am very talkative and cordial at lunch and supper.

Larry is out of town, and Paige calls with problems only a young adult can experience, so we talk. It is great to feel needed again. No advice is given. We just talk about feelings. I enjoy being a part of her life again. I enjoy the ability to communicate with her.

Julie, the 18-year-old I shared a room with on the neurology floor, is on the rehabilitation unit also. She is awake from her coma most of the time now and is in speech therapy too. Julie is a fighter.

I have seen young men in wheelchairs wearing body casts, the anger and frustration of other stroke survivors and quadriplegics. I have watched them when they were discouraged. I have seen their small grins as they advance in their rehabilitation. I have so many things to be thankful for. My mother used to say, "You feel bad because you have no shoes, until you see someone who has no feet." I understand the power in those words now.

July 23, 1992 Larry and Betty Rae visit during physical therapy. Sarah encourages me to work harder than ever for my audience. The therapist and I go outside to test uneven walking surfaces. We practice going up and down stairs, sidestepping, and finally Sarah demands, "Get on the floor, Cleo. If you fall at home, how are you going to get up alone?"

I bend my knee until I can catch the ground with my right hand, proceed to pull myself over to the nearest stable object, and pull myself off the ground using my right side. I do not indicate to anyone how tremendously difficult this task is to perform. I feel sad about performing in front of Larry and Betty Rae as though I am a trained seal doing tricks. I try to understand that Sarah's only goal for me is self-sufficiency. I want to show Larry not to be afraid of me. Maybe the performance is the only way to show him that I will not break.

I am on a self-medication record. I ask for my medicine at the correct times and record them on a daily sheet. I have to know what each medication is for and the correct dosage. Let's see; blood thinner, antiseizure medication, thyroid and hormone medicine, antiulcer medication, and acetaminophen when needed. I must remember how and when to take my medication in order to go home.

There is talk of a pass for me for this weekend. I must be patient and ask the doctor. I cannot afford to get my hopes up. A partner of my neurologist let the cat out of the bag. I am going home on a pass Saturday afternoon until Sunday evening. If all goes well, and I am sure that it will, I will be discharged on Tuesday. I begin to count the days. I will work very hard. Thank you, Lord!

Experiences in the Hospital and Heart Surgery

CLEO HAS DRAMATICALLY DOCUMENTED what it is like to be a patient in the hospital. The gamut of human emotions spills out of the account. She elaborates on the patient as an observer of life in a very special colony peopled with individuals with all sorts of illnesses and problems. This colony is very different from home and life outside a hospital. Death hangs in the air. Fear is almost a normal companion. Tests are ordeals—often uncomfortable, sometimes painful, and often frightening. Reassurance and a good advanced briefing about the indications for the tests and their nature help allay anxiety but of course do not remove the worry and fear. Everyone in attendance should understand this and do their part. Cleo's surgeon was very helpful when he spoke to her as the anesthesia took hold. After all, for patients this is a new, uncharted course; no one knows how it will turn out.

Surgery is especially frightening. Some patients are afraid of pain and welcome the anesthetic. Others recognize that some individuals who are put to sleep do not awaken. Cleo clearly felt the need before the surgery to say goodbye and thank those closest to her—her husband and her children. Even the most medically naive individual knows that the heart is vital for survival and fears heart manipulation.

Heart surgery has come a long way. Skilled anesthesiologists, expert heart surgeon specialists, and modern technology now often make heart surgery feasible even in the very old and the very sick. The frequencies of serious complications and death have become quite low in experienced hands. Still, physicians continue to explore new ways of achieving the same ends without surgery. Since Cleo's operation, a new device that can be threaded through the body's veins into the heart to close atrial septal defects and patent foramen ovales is now being tried in suitable patients. In the new procedure, specialized cardiologists skillfully direct heart catheters into the arteries feeding the heart to deliver drugs and dilate the narrowed, diseased arteries. The development of cardiac surgical teams with extensive experience and skill has led to fewer serious complications and deaths despite the fact that sicker patients are being operated on than in the past.

Intensive Care Units, Doctoring, and Continuity of Care

INTENSIVE CARE UNITS (ICUs) are especially scary places. Some patients in the unit are desperately ill. Nursing and patient care routines are conducted twenty-four hours a day, and the lights usually remain on. The bustle and continuity of activities, and the sedatives and medicines administered to blunt pain, make the experience a surreal blur for most patients.

Cleo remarks here on her experience with having an unfamiliar thoracic surgeon unexpectedly appear and perform an extremely unpleasant procedure. An often unemphasized aspect of medical care is the great importance of continuity. Getting to know and trust others takes time. All of us have different styles, mannerisms, and ways of communicating. Working with the same doctors, nurses, and staff is very important. Of course, doctors can't be there twenty-four hours a day, every day. But big procedures and discussions can usually be planned, except in unexpected emergencies to be handled by the patient's regular caregivers. Recently, after much public discussion about whether long hours lower doctors' effectiveness, rules were issued to limit the hours and times that doctors in training can work. Now we need to remember the importance of continuity of care, and set up ways of having someone available who knows the patient and his or her problems.

Cleo has described in her journal many different doctor and nurse encounters; and some make me cringe. There is much more to doctoring than scientific, technical knowledge and skill. Communication is critical and must be thorough, patient, timely, thoughtful, accurate, honest, and caring. Some physicians and health care providers are never able to master these skills that are often lumped together as "bedside manner."

Nurses

Every patient who has had any serious illness knows the importance of nurses. They are the bedrock of the hospital system. Without skilled nursing, medical care becomes fragmented and inhumane. Some critically ill, paralyzed but alert patients on respirators have told me afterward how totally helpless they felt and how they were always aware that their fate hung completely in the hands of the nurses. Failure to suction well or failure of the respirator to work could prove fatal if the nurses weren't skilled and vigilant. Caring, attentive nurses, as the traditional angels of mercy, can have great consoling and therapeutic effects on their patients. Likewise, inconsiderate, inattentive, and callous nurses can create very negative opinions and a threatening hospital milieu. Other nurses fall somewhere in between. Gloria, whom Cleo calls "the nurse from hell," was going by the book (including, as Cleo will eventually learn, reading her patient's notes from the trash can), but she didn't notice how closely her patient was watching nor how anxious she was feeling.

Stroke Treatment

The answer to Cleo's strokes was heart surgery, but doctors can use a variety of treatments for stroke patients. Treatment of a patient with a stroke depends very much on the type of stroke (ischemic or hemorrhagic) and the specific causes of that stroke in the individual patient.

In treating patients with an acute ischemic stroke, doctors first use two general strategies: they try to quickly bring more blood to the regions that are lacking blood flow, and they give medicines to reduce the chances of blood clot formation and extension. The tactic of bringing more blood to the threatened ischemic region is often called reperfusion because it entails restoring blood flow to regions of the brain that have recently been deprived of their normal blood supply.

But then we must deal with the blockage in the artery that caused the ischemia, and here we have a choice of strategies: 1) cut the artery open and remove the obstruction, 2) thread a tiny instrument into the artery to attempt to unblock it mechanically, 3) use a drug to dissolve the materials blocking the artery, or 4) create a detour around the blockage by attaching another piece of artery to the segment of the artery before the blocked area and attaching it beyond the region of blockage.

What we do depends not only on the patient's condition but on what artery is blocked and where it is. When the segment of blocked artery is able to be exposed, surgeons can operate directly on the artery to unblock it. This procedure is called an endarterectomy. Most arteries that commonly become blocked within the head are not readily accessible for surgery, so endarterectomy is used almost entirely for blockage within the neck. Surgeons place a temporary clamp above and below the diseased segment. They then open the segment and remove the inner core of plaque and thrombus; after the arterial segment has been cleaned out, the vessel is sewed back up and the clamps are removed. The process takes a half hour or less in most cases. Endarterectomies are usually feasible when the artery is severely narrowed but not totally blocked by a thrombus. But when a thrombus forms, it often extends far beyond its origin, making it impossible to open the artery surgically.

Alternatively, specialists can use an instrument inserted into an artery to dilate and open it. This process, usually called angioplasty, can be used to open arteries in the neck and the head, especially those arteries that are not readily accessible for surgery. Sometimes doctors use a stent to keep the artery open.

Thrombolysis is the term used to describe chemical dissolution of clots that are blocking arteries. The most common thrombolytic drug now used is tissue plasminogen activator (t-PA), which can be given either intravenously or by having a specialist place a catheter within the blocked artery to deliver the drug directly to the clot. To be effective, t-PA must be given rather soon after stroke develops.

If blocked arteries cannot be repaired directly, surgeons can reconstruct the blood flow channels by using various techniques. They can reimplant the distal end of the artery (the part toward the head) into another artery. They can also bring another artery into a position next to the blocked artery and sew the two vessels together. They can take a loop of another vessel from a distant site and use it to detour around the blocked artery.

Beyond these techniques, it is important, in general, to try to maximize blood flow to the head. When an artery is blocked, flow to the ischemic zone and the surrounding tissues comes from other arteries (called collateral channels) that ordinarily do not directly supply that zone. Ischemic brain tissue gives off chemicals that encourage ingrowth of collateral blood vessels to help take up the blood flow slack. Doctors may try to boost this process by maintaining or slightly increasing blood pressure, increasing the amount of body fluids, and decreasing the viscosity of the blood.

In treating brain embolism, some causes within the heart can be repaired— as in the case of Cleo's atrial septal defect—to prevent further embolism. When emboli break off from plaques within the aorta and the main arteries in the neck, these vessels can also be repaired. Another strategy commonly used in patients with either atherosclerotic vessel narrowing or embolism is to use various drugs that reduce formation of clots. These drugs are called anticoagulants. The most common drugs used are heparin (including low-molecular-weight heparin) and warfarin (coumadin). They are often mistakenly referred to as "blood thinners," but they do not affect blood viscosity. They simply decrease the tendency of blood to clot.

Drugs called antiplatelet agents are used to decrease the tendency of platelets within the blood to stick to plaques and rough surfaces. When platelets are activated by contact with diseased blood vessel and heart linings, they stick together and to the vessel wall, forming, with fibrin, so-called white clots. These platelet-fibrin clots can break off and block arteries within the brain. In addition, these white clots stimulate factors within the blood that act to develop red clots on top of the small white clots. Doctors prescribe a number of substances that decrease the formation of white clots. The most common ones currently in use are aspirin, ticlopidine (Ticlid), clopidogrel (Plavix), and a combination of aspirin and delayed-release dipyridamole (Aggrenox). Many substances used for pain and inflammation, such as ibuprofen and indomethacin (so-called NSAIDs), also have antiplatelet activity. Some of the drugs used during acute stroke, such as anticoagulants and antiplatelet agents, are also used to prevent reoccurrences.

Compared with the rather complex treatment of the different aspects of brain ischemia, treatment of brain hemorrhage is more simple and direct. Aneurysms that cause sudden bleeding around the brain (subarachnoid hemorrhage) can be clipped at surgery to prevent a second episode of bleeding. At times, aneurysms and vascular malformations can be obliterated by catheters placed within the feeding arteries; various substances can be delivered through catheters to correct vascular abnormalities. Vascular malformations can also sometimes be removed surgically or can be obliterated by radiation. When brain hemorrhages are large and threaten life, it may be possible to drain them by surgery. Reduction of blood pressure and reversal of any tendency toward excess bleeding can also be used to try to limit further hemorrhaging into the brain.

Newer medical, surgical, and radiological techniques have been developed over the past two decades that have increased the ability of doctors to treat patients with brain hemorrhages and brain ischemia. Better diagnostic technologies, such as MRI and CT, have made it much easier to precisely locate abnormalities and have greatly enhanced treatment.

Another strategy on the horizon to try to prevent or at least diminish brain ischemia is to give various drugs, often referred to as neuroprotective agents, that reduce the vulnerability of the tissues to ischemia. This would be like sprinkling something on the grass that would somehow make it more resistant to death in a drought. Neuroprotection is a very new idea, and many drugs are being tried in treatment trials. These drugs could be given soon after the first stroke symptoms and might give treating doctors more time to accomplish reperfusion and so preserve ischemic brain tissue.

Self-Image and the Brain's Ability to Appreciate Disabilities

Six weeks after her stroke, notwithstanding pain, setbacks, surgery, and surprises—both pleasant and unpleasant—Cleo has improved considerably. Some of her journal entries tell us about her problems with speech, memory, vision, and feeling on her left side. She clearly is concerned about her self-image after the stroke. She tells us of difficulty with simple things like tying her shoes, and of falling and struggling to get up. Early on, some patients with strokes—especially strokes involving the right side of the brain—often do not fully realize their handicaps and deficits. They also may not be able to see themselves as others see them. Later, return to daily activities and responsibilities shows them what they can and cannot do as they did before.

Both psychological and neurological changes follow strokes. Strokes often create deficits easily visible to everyone, even during casual encounters. Something in an abnormal facial appearance, use of the limbs, walking, or speech are often quite obvious to all observers. Cleo tells me that she worries about a limp when she walks. Neurological patients often wear their illness on the outside, while most diseases of the other organs of the body are usually not readily visible to others. Since their bodies seem changed, it is quite natural and understandable that patients will worry whether the changes will prove acceptable and tolerable to their significant others. This becomes especially problematic when the caregivers must alter their own activities in important ways to help the patient. Will they be willing and able to handle the new responsibility? Will the caregivers pitch in happily, or will the patient feel like an unwanted burden? Spouses, children, and others must understand that patients naturally have these worries, and try to calm and reassure them whenever possible.

It may be surprising, but behavioral changes due to stroke are usually in the other direction. That is, patients do not fully comprehend the nature and severity of their deficits and are inappropriately unconcerned. The right cerebral hemisphere is probably mostly responsible for both awareness of deficits and emotive responses to them. Some individuals with large right cerebral hemisphere strokes, who have left limb paralysis and do not pay attention to their left visual field, may completely deny that anything is wrong with them. This lack of awareness of the deficit is called anosognosia (meaning lack of perception). I recall rather vividly being called to the emergency room of my hospital at 3 A.M. by a perplexed and desperate husband and house physician. A woman had very suddenly developed complete paralysis of her left limbs and was brought to the hospital by her husband despite her resistance. She absolutely refused to be admitted to the hospital, declaring that nothing was the matter with her. Neither the husband nor any of the emergency room personnel had been able to convince her otherwise, so I was summoned. Indeed, she had complete left-sided paralysis, loss of sensation in her left limbs and body, and a lack of response to visual stimuli on her left. After the examination she said, "You see, nothing is wrong with me." Rather than argue, I got her car keys from her husband and gave them to her. I told her if she could walk to her car and drive home, then she could go. She tried to get up and slipped to the floor, unable to rise. She accused me of tripping her but said that now she must stay because of her fall. Not until the second week of hospitalization did she realize that anything was wrong with her limbs, feeling, or vision.

This is an extreme example, but caregivers do not always understand that stroke patients may not fully realize what's wrong with them or what they can and cannot do. It only becomes apparent when they have to perform a task. Cleo will become more aware of her problems as she does more things, especially when she is faced with more activities and responsibilities when she goes home. She will get a preview of this now, going home on her weekend pass.

July 24, 1992   Although I am closer to going home, I continue to struggle with the four-letter word TIME. I want the process to hurry up, but healing is slow, it will take time.

I am apprehensive about the ride home tomorrow afternoon. I am anxious about what it will be like to be home. I know that it will

be extremely difficult for me to return to the hospital Sunday evening.
God grant me patience. God grant me peace. God grant me your healing power.

July 25, 1992 I awake early this morning. I slowly dress myself,
put some lipstick on, and watch the clock for the
time Larry will come and the joyful weekend will begin. I am excited
yet genuinely concerned about what the weekend will bring. I am finally going home.

The car ride home is filled with mixed emotions. As we get closer
to home, I get even more excited, yet wonder what it will be like. Seven
weeks is too long for me to be away. The neighborhood, the familiar
houses, finally we arrive at our home.

The "For Sale" sign stands ominously in our front lawn. I hate it,
but I understand. We cannot go from a two-income family to one
income with mounting medical bills and expect our lives to
be unchanged.

As I carefully move from the car, Larry helps me up our walk to the
front door step. I look down to watch each footstep very carefully.
The sidewalk catches my attention. There in the concrete are the hand-
prints of our children and the year we built the house. I remember
the day they placed their hands there, wrote their names, and noted
the year above their prints. How excited we were to move in to our
dream house.

Paige and Mark hold back their long conversations and settle for
small talk over the weekend, not knowing exactly what to say. The
only one that seems normal is our dog, a miniature schnauzer named

Cindy. She welcomes me with barks and jumps and stays by my side most of the time.

I walk from one room to another, looking, touching, and remembering. Everything is the same, at least for now. I seem to see the ghost of myself running upstairs and down, calling the children, laughing with Larry, seeing myself as I used to be. I do not want to be a burden on my husband or children.

Claire and Don are here to welcome me home. As we sit outside beside the umbrella table the words do not come easily for me. I still need speech therapy and this afternoon I do not feel like trying to find the correct syntax.

It is suppertime and I try to assist this effort by setting the table. "I'll do it! Things have changed. For a while you're a guest—enjoy it!" Larry says emphatically.

This is my home too! At the supper table I look at everyone as they sit in their chairs, bow their heads for the blessing, and pass the serving dishes. It is too much for me. I excuse myself, retreat to the bathroom, and cry. I missed that family scenario so much, and it is overwhelmingly wonderful to be able to be part of it again. For my family's sake I will try harder.

I watch television with Larry. He puts his arm around me and I fall asleep. When I awake I talk with the family, take my medicine, and Larry helps me climb the stairs to our bedroom. It has been a busy and frustrating day. In my home I want to be whole again. It has been a beautiful day, because I am finally home. I have made it through the worst of times, God and I. Please continue, Dear Lord, to help the family and give us patience.

July 26, 1992 Larry and I awake early. I take my medicine. Larry goes to the gym to work out. Today will be busy with my sister-in-law and niece coming to visit.

Larry and I go next door to visit the neighbors. My sister-in-law and niece join us for a day in the sunshine.

I have a hair appointment scheduled for this afternoon. Larry is insistent that I will look much better with my hair cut short and wispy. My friend has made the hair appointment and provides transportation across the city. I am extremely tired when we return and take a nap before dinner.

It will take time to assimilate into my family again. Larry wants to run the house and have me concentrate on getting better.

Larry bought me a new pair of walking shoes. He said the running shoes I had for so long had outgrown their usefulness. He will gladly buy me a new pair next year if I can run by then.

The pain in my left side is excruciating all weekend, as if I am connected to an electrical current. My face, ear, shoulder, arm, and down my left side to my toes feel cold and tremendously sensitive. As I walk, my left foot feels like a heavy object. We had it worked out in the hospital. I would take two pain pills about a half hour before therapy sessions, and then we could work without pain for at least two hours. I do not know how to do this at home. How much pain medication can I take? When? The atmosphere at home is different from the hospital. I will have to learn to care for myself now. I am confused. The neurologist had told me that the pain was a good sign in my recovery because it meant that the brain was making new connections. Do I have to deal with this until I completely heal?

The ride back to the hospital is eerie. Larry and I haven't said a word to each other. I feel as though I am being punished for some unspeakable crime. I do not want to return to the hospital. However, it is probably best for a couple of days as my endurance and strength are not good yet. The nurses ask me about the weekend. Before I can answer, Larry states that about every two to three hours I needed to take a nap. I curse his opinion of my time home, even though it is true. What is he trying to do? I want to be discharged in two days! I have worked so hard in the hospital. The therapy schedule I am on does not permit me to take naps or long, leisurely rest periods.

It was good to be home. Things have changed. No one is more aware of that than I. However, I must be able to meet this challenge. We will be fine.

Pain in Body Parts That Have Reduced Sensation

AS CLEO IS RECOVERING from her strokes and is undergoing a rather vigorous therapy program, pain in her left face, shoulder, and limbs has become a vexing problem. She describes the pain as "icy cold." Recall that her first stroke involved places in the brain that receive signals from the body (somatosensory nerve cells and tracts). Cleo's stroke involved somatosensory areas in the right thalamus and white matter in the right temporo-occipital lobes. She had a severe loss of sensation in her left face, trunk, and limbs, as she explained it, just as if someone had drawn a line down her middle and she had lost all feeling on the left side of that line. Although it may seem paradoxical, pain in areas of the body that have lost feeling is rather common and usually develops some time after the onset of sensory loss. The phenomenon has been referred to as anesthesia dolorosa, meaning, literally, "painful anesthesia."

The somatosensory system has two different types of sensibilities. One type, related to coarse sensations such as pain and thermal sensation, is called *protopathic.* The other category of sensations, describing sensations that are fine and precise and relate to touch, joint position sense, and the ability to detect vibration on the skin and bony prominences, is called *epicritic.* When all nerve pathways are intact, the epicritic, or finer, sensibilities predominate. When these more precise nerve pathways are impaired and parts of the body are anesthetic or fine touch perception is reduced, then all stimuli seem to evoke only coarse, unpleasant sensations that patients often describe as very hot or very cold or stabbing.

You can think of this as being like a finely tuned radio that has been dropped into water. Wetting the delicate fine wires impairs the performance, and fine music is turned into mostly static. The louder you turn up the volume, the more the static grates on the ears. Stimulating an anesthetic arm similarly stirs up unpleasant sensations that become more and more intolerable with increasing stimulation. Working with the arm during therapy and sleeping with unrecognized pressure on the arm can provoke severe pain. Unlike ordinary pain that arises from injured local tissues, this pain is "central" because it involves abnormal function of nervous system tracts, not any injury in the area where the pain is perceived.

Doctors often refer to pain from something wrong in the central nervous system as "thalamic" or "central" pain because most often the problem is in the thalamus, the main way station for receiving sensory input. Recall that Cleo's stroke damage included the thalamus. Sometimes when doctors explain the nature of the pain and advise the patient to persist despite it, the patient can learn to ignore the unpleasant sensations. Ordinary pain medicines are not very helpful. Some drugs also used as anticonvulsants, such as phenytoin sodium (Dilantin), carbamazepine (Tegretol), and gabapentin (Neurontin), as well as small doses of amitriptyline (Elavil), may be effective in diminishing the pain and the response to it.

Recovery from Stroke

CLEO'S WEEKEND AT HOME was a test, and it demonstrated that she is definitely on the mend. The fact is, the great majority of stroke patients get better. Some improve so much that they return to normal or near-normal daily life. There are three general explanations for improvement. First of all, some of the injury—ischemia or swelling—may be reversible and the tissue injury may heal during the days or weeks after the stroke. Second, very few if any brain functions are completely confined to one site; when one region is injured, other areas can take over. This takes time and is clearly influenced by activity. Talking to and with aphasic patients stimulates uninjured regions to increase their language capabilities. And third, people adapt to the deficits. That is, they learn to do things differently than before.

Individuals vary widely in their ability to return to normal. Considerations that influence recovery of function can be divided into four groups:

1. *Disease related.* Hemorrhages and infarcts have different timing and extent of recovery. Patients with hemorrhages get better more slowly but often improve more than patients with infarcts of similar size and location.

2. *Anatomy related.* As I have emphasized, different parts of the brain do different things. The location of the stroke is more important than the cause in determining the types of deficits and handicaps a person may have.

3. *Individual related.* We all know some healthy individuals who are motivated, well organized, determined, and successful, while we would describe others with opposite adjectives. The attributes, capabilities, and failings of individuals are even more important after a stroke, because they correlate closely with the patient's ability to overcome adversity. Someone who has not been able to hold a job before a stroke is extremely unlikely to return successfully to the workplace after a stroke. Faith and determination are also extremely important in whether or not someone will overcome handicaps and be able to bounce back.

4. *Environment related.* Personal, interpersonal, social, and economic resources are extremely important. Perhaps the key element is the presence of one or more close, caring people who will give the stroke patient help, support, and encouragement. Such ordinary considerations as having easy transportation available, living on one floor, especially the first floor, the presence of an elevator in an apartment building where the patient lives on an upper floor, accessible shopping, and handicap precautions and facilities can all make a world of difference in what stroke survivors will be able to accomplish. Money to pay for adequate equipment and help is also an important consideration. Cleo often mentions financial worries.

Having returned without a setback from her weekend at home, Cleo will spend only one more day in the rehabilitation hospital, and then, as her doctor promised, she will be discharged to return home for good.

CHAPTER 5

Homecoming

July 27, 1992 The neurologist visits me in physical therapy today and says that I am ready to go home tomorrow. He tells me that I must go into the community hospital every Tuesday and Thursday for blood levels to be checked to regulate the heart medication and to be followed by my physician. He will also plan outpatient therapy and transportation to and from the smaller hospital. I am overjoyed!

Thank you, Lord, for the qualified and professional medical staff and therapists I have had work with me. Without their help I would not be as far as I am. I can slowly walk by myself. I can speak clearly and distinctly, although sometimes I get high-pitched or monotone. I do need some work on my voice tone, and I have trouble hearing. My left arm moves, but not smoothly. I continue to need fine motor development, since my fingers don't react. Considering where I have come from, I am doing very well.

I remembered what I asked the Lord: "Please Lord, just let me see the milestones in our children's lives—their growing years, graduations, marriages, and children." God answered my prayers. How can I ask for more?

July 28, 1992 It is imperative that I learn the names, dosages, and side effects of every medication I am taking before I am released from the rehabilitation hospital. I must learn why they are given, when to take them, and how much they cost.

Coumadin is an anticoagulant, or blood thinning medication. The neurologist prescribed this medication, along with daily blood level checks, after my first admission. I was taken off the medication prior to surgery. After the surgery I was placed back on Coumadin.

I had a hard time last night getting on top of the thalamic pain registering deep electric or burning stimulation to the entire left half of my heavy, awkward body. I keep telling myself that the deep nerves are returning. If I overdo any activity or don't take a pill at the first painful moment, I cannot get the pain under control. I cannot relate this to the nursing staff today. I am afraid they won't let me go home. I must try to be optimistic.

This morning, the sun is shining and it's a great day to go home. I have to keep all appointments for my daily therapies. I am nervous awaiting Larry's arrival. He finally arrives with Betty Rae during the supper hour. "Let's go!" I shout with enthusiasm within my mind. We pack, receive my discharge orders, prescriptions that need to be filled, and say good-bye to the wonderful staff.

Larry goes down to the hospital pharmacy to fill the prescriptions. When he returns his face is cold, angry, and his head shakes at me.

"Do you have any idea how much these prescriptions cost?"

"No," I answer.

"Close to three hundred dollars, and it's an out-of-pocket expense I wasn't prepared for."

I blame myself for putting such a financial strain on the family.

It seems like such a long ride home from one side of the city to the other. The four-lane freeway is jammed with cars during the rush hour. I am claustrophobic, but I know we are getting closer to home with every mile.

When I arrive at home, Paige and Mark are out with friends for the evening. I sit quietly in the family room, alone. The evening progresses very calmly. I do not want to upset Larry even if it means separating in different rooms to avoid each other.

July 29, 1992   I awake abruptly at 4 A.M.; I have wet the bed. Larry is angry and swearing. He will not let me help change the linen. As he washes the sheets and places the clean linen on the bed he continues using four-letter words. I feel ashamed. I do not feel like his wife and lover but like an invalid he now has the burden to care for. Not only does he have the responsibility of trying to sell the house, paying for everything without my income, caring for the children, cleaning, meals, and shopping; he now has a person who needs to be diapered at night. This is not going to be easy.

In his own way, Larry tries to apologize for his actions: "Well, you make one hell'uva alarm clock!" I cannot laugh. I know the pressure he is under.

I make my appointments for outpatient rehabilitation and schedule transportation by myself this morning. This helps my ego.

I feel invisible. The children intermittently file through the house to get this or that. I love to be able to see them again whenever I want.

The real world is entering the picture now, and the children seem more relaxed about their mother and more intent on themselves. The way it should be.

Tonight I got a big hug from Betty Rae. She wants to help me as I dress for bed. I ask her if she can find my pink pajamas with the lacy collar. It is hard for me to tell her that I need to dress myself, even if it looks awkward and painstakingly slow.

Bladder and Bowel Control and Stroke

CLEO HAS HAD A very disconcerting experience—wetting the bed. The experience was a bad one for both her and her husband. Although we all accept the fact that children cannot control their urine or bowels, somehow the same problem in adults is very difficult to tolerate, especially in our spouses. Perhaps one of the reasons relates to sexual activity. Since sex involves the same anatomical regions, the potential specter of incontinence psychologically changes the sexual experience, introducing the possibility of soiling during intercourse or other sexual activity.

The nerve centers that control urination, defecation, and sexual genital functions are located in the lower spinal cord, in a region comparable to the region that controls the tails of animals—the sacral region. Many of the functions are reflex. Filling of the bladder creates the sensation of the need to urinate, and filling of the bowel creates a sensation that stimulates bowel evacuation. Centers in the upper spinal cord and brain also affect these functions.

In the brain's frontal lobes, next to the motor regions that control the legs on each side, there are centers that relate to voluntary control of urination and defecation. When the urge to urinate occurs, these brain centers maintain voluntary control of the urinary sphincters. By age 2 or 3, humans have learned to inhibit relaxation of their external urinary sphincters, delaying release of urine until they reach a suitable place to empty their bladders. In patients with strokes and high spinal cord injuries, the control over urinary release is often affected, and the reflex-induced bladder contractions are hyperactive, just as the limb reflexes are exaggerated. As a result, once the urge occurs, they must quickly empty their bladders or risk being incontinent. Often they cannot control or feel the need to release urine, especially when they are asleep. The voluntary control fibers are on both sides of the brain. A stroke affecting one side of the brain, or the descending fibers on one side, usually causes only temporary loss of control of urination because the other side will take over control with time.

Similarly, brain centers control the external anal sphincters. Normally, adults can inhibit relaxation of these sphincters until they reach a suitable place to defecate. Strokes can affect the ability to control bowel release, especially if the bowels are loose. Bowel incontinence is a less common and less severe problem than urinary incontinence, since reflex functions are often sufficient to maintain bowel continence, especially with some training.

July 30, 1992 I can't sleep. My left arm feels like it is tightening in the elbow. I cannot allow that to happen.

The first rehabilitation session is scheduled for August 3. It will be three days a week. I want to take afternoon naps, just like preschoolers. I listen to my body more. I know when I need to rest, how much to eat, and when I should exercise.

I simply observe. I feel as if I am no longer a part of the team. I am benched. On this team you either play your position or get out of the game. I have no other choice but to leave the game, as I feel I am no longer a contributing member. I am a loser, a deficit, a flighty reactionary because of my hypersensitivity to things around me. I don't want to touch anything with my left hand, as everything feels as if it will hurt me. I can't even feel the moisture of water, the roughness of upholstery, or the smoothness of my girls' faces. I no longer cook, clean, and provide income as I once did. No longer do I move fluidly and chime in on conversations with a thought or an idea that expresses part of my individuality. I am isolated and lonely in a house full of the closest people in my life and I can't remember really knowing them at the core of my being like I should. I'm frightened.

Mark is usually with his friends or an occasional "Are you okay, Mom?" is all I hear. It is his way of reassuring me. Mark is quiet, a spectator of life, and it is his unsaid words that mean the most, the glance, the smile, and the touch on the shoulder.

Paige is talkative and spends most of her time in the bathroom getting ready to go somewhere with someone. I listen to her ranting and raving, "Well, Mom, you haven't heard me bitch in seven weeks!" We laugh.

Betty Rae is my helper. She always wants to fix me something to eat, sit by me, and hug me. I love every minute of it. She leafs through a photo album with me by her side. I look at the pictures and know it is me with the family. I stare at pictures that were taken at different places and spaces in our lives, but I can't focus for more than a few seconds and then my mind is on to something else—the bark of the dog, the quick motion of Betty Rae's hand, or her voice. I can't remember exactly when the pictures were taken or even being a part of them. I don't remember the person I was. Everything is new, and I don't want to tell my family that I can't remember how old they are—only their birth dates. I can't remember holding Betty Rae when she was a baby and I can't remember how old I am today, the date, the month, or at times, even the year. Betty Rae needs her mother. We will have to work on this together. Life must get back to normal, though I have no idea what "normal" is anymore.

July 31, 1992 No one can change the past. I must learn to let go. I will keep focused on the present and future. It is not what I write about that will change me or give me determination. However, it is how I deal with the things that have happened that will be my making or breaking. I am a survivor. With God's help, I will see my children grow. I have an enormous potential within me to love and to be loved, if only I can hang on to that thought for today. Life is a menagerie of events.

I can pity myself into such a deep depression that I may never see the happiness of another sunrise. I cannot allow myself the emptiness. My father died in 1966, at the age of 47, from a heart attack. Today, medical advancements saved my life. My mother died four years later, in 1970, in a car accident. Seat belts are mandatory now; maybe it would have saved her life. My older sister has difficulties related to brain injuries caused by a car accident that resulted in lack of oxygen to the brain. She and I have similar speech and movement difficulties, although her injury involved the left side of her brain and right side of her body. We have laughed and joked about tying our affected sides together and making a run for it. We have given each other permission to laugh. My brother died of AIDS. I hate the disease AIDS, as I hate stroke, cancer, and heart disease, and I empathize with people who fight to overcome these devastating diseases.

If given the opportunity to pass on to the next generation what I have learned from this tragedy, it would be survival. To take a deep breath and say, "What have I learned from this? Where do I go from here?" I hope that my children can pick up the pieces and go on, knowing that they have been somehow strengthened by this experience. Stroke plays an enormous role in my life. It is part of me. However, I will not allow it to define and dominate my existence.

The Need to Look Forward

CLEO IN THESE LAST PARAGRAPHS has hit on one of the most important keys to recovery—the need to look to the future and not to dwell excessively on the present or past. All too often, stroke patients continue to dwell on their misfortunes and the calamity that has befallen them. They become obsessed with the fact that they are not exactly as they were before. The change—their stroke—has come suddenly, but actually in life none of us are as we were decades ago. The change is so slow and imperceptible that we seldom recognize it. As our middle years come on, we all realize that we cannot perform activities that were routine when we were teenagers and young adults. Successful people do not dwell on this loss of capabilities. They emphasize different directions. With time comes experience and changes in goals, directions, and activities. Our lives are a series of changing passages.

Successful recovering patients are determined to return to normal activities despite any handicaps. I, as their physician, emphasize to patients that they may not need to return to perfect normalcy to perform everyday tasks. For example, people can learn to eat with one hand. Some people learn to drive well despite arm and leg weakness on one side of the body. Patients who look forward and are determined to get on, no matter what, usually pass smoothly through the return from "patient" back to "person." Those who dwell on their handicaps trap themselves in the idea that they must be exactly as they were before the stroke. They remain eternal patients and often forget to live.

Many patients become depressed after a stroke. The depression usually starts after patients begin to realize what has happened to them and how the stroke has affected their lives. Some depression may also be associated with brain tissue damage and healing. In either case, when depression develops, most often it responds to antidepressant medicines and compassion.

August 13, 1992   I received postcards and letters from our friends in Montana throughout my hospitalization; now they are coming to visit. Our home is certainly large enough for overnight guests. The bedroom and bathroom on the lower level of our home will provide for their comfort. We will have a dinner party with another couple that knows them quite well.

I help as much as I can to prepare dinner and make sure their bed has fresh linen. However, Larry does most of the work, deciding to barbeque hamburgers on the gas grill out on the deck. I enjoy laughing and joking with my guests, and for a short while I try to forget about the stroke. It is past midnight when I retire with Larry.

At around 3 A.M., I begin a grand mal seizure. Suddenly, the bedroom light is on. Larry grabs my shoulders. I hear him shouting at me, but I am too tired to respond. He keeps shaking me. My eyes are too heavy to open. He grabs my ears and raises me off the bed. He is extremely angry and swearing. I can hear him but I cannot respond. I try but the words just slur. Our friend who is visiting is an emergency room nurse who can handle this situation, but Larry does not involve the guests. His teeth are clenched and I can hear him muttering, "I've had enough of your gibberish." He clamps his hand over my mouth and nose. My eyes are too heavy to open. When he releases me, I attempt to sit up and catch my breath. He continues to swear as I fall back on the bed: "You fucked up everything. Wake up now!" He lifts me off the bed by my ears again and spits in my face. I feel the warm expectorant run down my cheek. He places his hand again over my mouth and nose and says, "It would be so easy." I can feel my heart beating strong and hard as if it were in my throat. I am too tired to resist and slump down to sleep. With a spurt of energy I take a breath and try to fend him off with my right arm while trying to communicate. Suddenly, he grabs my neck, "Wake up! You're fucking me up. You fucked everything up." I can see blurry vision off to my right. The digital clock by our bed indicates the hour of 3 A.M. in bright red numbers.

"Fine," the word flows from my mouth. "I'm fine. Let me sleep!"

"I'll let you sleep all right, you bitch. How dare you die on me!"

Quietly, I lay listening to my heart beat against my chest as if it were going to explode. Calm down, body. Don't start another seizure. I don't think I can live through another one.

I ask for a glass of water. Larry's reply is, "Can you get up and walk?" I didn't want to attempt walking now. I did not wet the bed, which is a good sign. Now I am terrified of everything, even my husband! Don't cry. Don't let him know you could hear him. Don't let him know how he hurt me.

In the morning he kisses me good-bye. He telephones me throughout the day. "How are you, honey? Maybe you should cancel therapy today and sleep."

I telephone my neurologist. "We have to talk. My family is terrified of my seizures. They don't know what to do when they occur." All family members are asked to attend a meeting with the doctor. Larry is too busy working. Questions from the children come in rapid succession at the meeting. "We want to get these seizures under control. Maybe we need to change your medication," the neurologist comments.

I want to tell him what happened. Instead, I tell the hospital chaplain and read the journal that documents every detail. He listens, but we both know the choice. I will be sent to a nursing home if I can't make it work. Can I risk staying in this situation? What about the children? Will time heal this wound?

When anger is not released it turns into rage. Larry and I begin to avoid each other as much as possible. We do not speak to each other about the incident. He seems to be constantly on the telephone to clients. When I try to speak to him, he walks away as if I am part of the furniture. There is no eye contact. I go to bed early and alone now.

Within a few weeks, I begin to speak of the fear. "Larry, I can hear you but I can't move or even open my eyes when I have a seizure." He sits in bed with his face covered by a book he is reading. However, he listens. "That night we had guests. I was terrified of you." He interrupts and says, "Am I not entitled to a nervous breakdown?"

I am silent. Over the years I had learned never to voice my opinion or get into a verbal confrontation with him. With my speech so monotone and gravelly, and words mixed up so much, I have to pick it out of my mind and bring it to my mouth. I didn't answer. But I thought, "You are not entitled to take that anger out on me. I will not live in fear."

September 5, 1992   I need Sarah. I need my therapist. After eight weeks of seeing her twice a day, exercising with her, walking with her, talking with her, we have become friends. I walk with a definite limp now. My left arm moves but it hurts most of the time. I am quiet and agreeable with just about everything. I am definitely backsliding. I can't move my arm, and it is bent at the elbow and lies across my stomach with my fingers frozen straight out. At times, I notice the long fingernails of my left hand. My limp is getting worse, and it's impossible to climb the stairs several times a day, so I usually decide to stay upstairs in the bedroom.

Mark ignores me. Betty Rae loves me. Oh, how she loves me. "Can I sleep with you for a while, Mom?" I rub her back gently with my right hand. We speak not a word but when she returns to her room for the night, she kisses me and says, "Goodnight, Mom, I love you." It is all worth it again.

Larry hasn't kissed me since I have been home. I know it is a love-hate, patient-caretaker relationship right now. He is afraid of me. At night, he watches television and then reads. He makes no room for me in his life. The tension is always there. Larry is going through the motions day by day. We don't communicate anymore.

Emotions seem to frighten this stoic man. He has always been my knight in shining armor, my comforter, and the sage of the family. He can find a solution to most any problem. He is strong and intelligent. But I am frightened for him now, and I cower in fear for what an explosive powder keg his life is right now. Who holds him when he cries? I must learn to forgive him and allow him to heal, too.

September 11, 1992 On Friday I attend Occupational Therapy as an outpatient. During an exercise of ripping paper into strips, I have another seizure. The therapists rushes me to the emergency room. After CT scans, blood tests, and rest, I am discharged into Larry's care.

Saturday I have another seizure and sleep straight through until Sunday night. During this time I don't eat or drink. I sleep, use the bathroom, and take my medicine. I am left alone and isolated. Everyone in the family goes about his or her daily activities.

I have a tremendous headache. I am nauseated and it is hard to swallow. I want to lie still. I cannot tolerate bright lights or loud noises. I cannot see images from my left eye, and my left ear is buzzing again.

Larry approaches my bedside. He hurt his hand while working on the car. I tell him, "Wash it well with salt and pepper," instead of saying "soap and water." The aphasia is back.

September 16, 1992 I have to get up today. I am feeling better and I know that if I do not move about, my muscles will constrict. I learned how to do basic hygiene in the hospital and now perform the tasks quite well at home. Now I am going to go for the gold and shave my legs. I know I shouldn't, since I am taking blood-thinning medication, but I have such excruciating pain when using the electric shaver on my legs that I'll have to adopt another style of accomplishing the task.

I draw a small amount of warm water in the deep tub, test the temperature with my right arm, get Larry's shaving cream and razor, and slowly climb in. I have to have a game plan for this procedure or I know I will surely drown. I am home alone. The tub is so deep, and turning about to use only the right hand is dangerous. I know my left side is of no help. Remember, don't cut yourself or you could bleed severely. After a few minutes of wiggling, whirling, and wrestling in the tub, I feel as if I have done an adequate job. I rinse.

Remembering what my mother had told me about babies drowning in a small amount of water if unattended, I want to scream, "What about me, Mom?" Is it the stroke that makes me miss her so deeply, or is it because I feel I can't perform even the smallest of tasks by myself? God, I miss her now.

Slowly I stand up, sit on the edge of the tub, and swing both legs around. I think it must be the same principle I learned in therapy for getting in and out of a car. It works. I slowly dry off and sit down to dress myself.

I feel excited about my new accomplishment. Maybe I can push myself a little further by doing some laundry. To my delight, I complete the task. Today will be that extra-special day.

I am going to try a new recipe for supper tonight and surprise my family when they return from school and work. I had better start early! It may take me all day. I try to peel potatoes but the first one flips down the garbage disposal. I can't hold it with my left hand. Maybe I should wait until Betty Rae returns from school to help me.

After supper is prepared, the table set, and the meal served, Betty Rae says the table prayer. Larry glares at me and says, "You like bland food, don't you?"

"It's a new recipe from your sister," I say, but maybe I shouldn't have tried it. The dicing of potatoes and onions cut my left finger a little, anyway. I had to say something as I watched him try to doctor it up with almost every spice we had in our cupboard.

Shortly after supper, he announces, "Who wants a Dairy Treat?" Larry knows the new medication I am on for seizures might make me hungry and that I had lost over thirty pounds. I don't need to return to my old habits. I am not extended an invitation for the evening ride.

We always had a ritual when it came time for bed. Watch the news and one half-hour show on television while in bed, read for awhile, then off to sleep. Tonight it's going to be different. After all, I shaved my legs! After the television shows are over, I put my head on his shoulder and try to surround him with my body ever so gently. He begins a nervous tittering and pecks me on the lips. He grabs his book like it is his saving grace. I quickly spring up to the side of the bed, put my glasses on, and go for my book. The book is just a disguise; so are my glasses. I still can't read well and if I try too hard, I get a fierce headache. I try to ease the uncomfortable situation: "It's all right, Larry, I understand." I can't look at him when I say it because in my heart I don't understand.

As I sleep, I have a dream that I am whole again, beautiful and sexy. For some strange reason, I dream I am in bed with Richard Gere.

He kisses me slowly, tenderly. His lips are warm and smooth. We kiss over and over again; we hold each other gently and talk. We talk about insignificant things like what we like to do. We talk about funny things like our most embarrassing moment. We laugh. Oh, God, in my dream, I laugh. We kiss, and talk, and laugh, all through the night. Then he says, "It will be morning soon. I need to go." I wake.

It is morning. I know immediately that I am not a "pretty woman" and no beautiful red dress is going to change it. Shit, my left side feels cold. It's time to enter reality. However, I keep thinking that Larry could have been my Richard Gere if he'd wanted to.

September 17, 1992 Today I bake four loaves of bread from scratch and make a large pan of caramel rolls. I am determined to use my left hand in some type of home therapy.

When I mention this to the children, their eyes seem to light up. Larry is extremely surprised.

At first, I have all the confidence in the world. I certainly have the time. I find the recipe in my favorite cookbook. I can't read it. The print is too small. I find a large magnifying glass that we inherited from my mother-in-law and begin again.

The ingredients are the next to be organized. I can't remember where we keep the flour or baking soda. When I find them, I can't remember which ingredient has already been placed in the bowl. I have to start over. I devise a check-off system to help me with this chore. As I begin to stir the ingredients, the bowl slips off the counter like a flying saucer. It lands in pieces on the kitchen floor. My left arm will not hold the bowl.

I start again. This time I set the bowl in a drawer. By closing the drawer against the bowl, it holds the bowl in place. Finally, the dough is ready to knead. This is the chance I've been waiting for. I want to punch something. Both fists, right then left, slowly smash the dough, and then repeat. No wonder the dough doesn't rise as well as it should. By using the dough as a therapeutic tool, I am developing strength, not kneading dough. With the dough in the rising stage, I look around the kitchen. Flour covers the venetian blinds, counters, and cabinets, and shards of glass from the first mixing bowl glimmer on the floor. Dough is splattered and dripping in big globs from the windows. Maybe if I close my eyes it will go away. Under my breath, I curse the stroke and my ineptness and angrily agree that I am not ready to perform such a task quite yet. Slowly cleaning up the mess, I carefully place the raised bread and caramel rolls in the oven. When the timer goes off, my job is finally complete.

Larry came home late last night and left early this morning. When he returns at the end of the day the family gathers in the kitchen and watches the celebration of my bread accomplishment. The caramel rolls look just like the picture in the recipe book but are as hard as rocks. Maybe the microwave will soften and warm them. The bread and caramel rolls are a silent success because, quite frankly, my family knows that I am trying, and the end product is not as important as the attempt. By the next morning every caramel roll is eaten and the bread is going fast. Delighted at my culinary talents, I throw something in the trash and spy the caramel rolls hidden under a mound of coffee grounds.

LARRY INVITED GUESTS for supper tonight and all I've prepared are bread and caramel rolls! I forgot he extended an invitation for supper. The bread has taken me so long to think through and complete that I've forgotten about the main course. He'll have to go to the store for something to put on the barbecue. The girls and I will prepare a salad.

As Larry leaves for the grocery store I say, "You forgot something" and give him a peck on the cheek and he blushes. I am happy.

By the time I go to bed, I have overdone again. My side feels as if it is in ice, and no matter what I do I can't warm it. I try not to cry. I muffle the sound so I won't disturb Larry. Without rolling over or turning on the light he says in a chastising way, "What's the matter?"

"I don't know. I'm going to have times like this, you know. I hate this stroke. What happened to me?" I say between sobs. Not a word, not a twinge from Larry's body. I feel so lonely. I finally fall asleep.

PAIGE. I am worried about Paige. She should have gone to her class today. College classes started yesterday. She told me that she had a counselor appointment due to missing her final exams the morning of my stroke. She also explained that she had completed her make-up work at the beginning of the summer.

Tuition is paid and she is scheduled to attend today. I know I have to stop worrying about her, but I will try my best to help her to achieve her goals.

I try to remember: What was it like to be 19? Somehow I can't help thinking about my past. I want her to have the childhood I was denied. In 1966, when I was 17, I drove my father to the hospital when he had a heart attack. He had come home from work about mid-morning, which was very unusual for him. I was getting ready to go to work at my summer job as an usherette at a local theater. My father lay down on the couch and said that his arm was hurting him but that he would be all right. Before I left I peeked in on him. He was barely conscious and I had trouble rousing him. His face was ashen gray. I was young and stupid and didn't think of calling an ambulance, just as Paige and I had forgotten to do the morning of the stroke. Where was my brain? But at 17, I assisted my father to the car and drove to the hospital. My mother, three sisters, and two brothers were up at our summer cabin thirty-five miles away and had no telephone. The physician told me that I had better gather the family as my father had suffered a massive heart attack. Because I only had my driver's permit and was merely a beginner driver, or because I was too stressed to accomplish the task, I got a friend to drive up to the lake. My father lived for six more months before he had another heart attack and died. He was 47 years old. These events are so vivid in my memory I feel as if I am there and I can't grasp the reality of my present family. I am disjointed. I can't remember ever being 19; a fleeting thought will not stay in my brain long enough. I perspire while trying to remember.

Deep in her soul, is the morning of the stroke hurting Paige as much as my father's trauma played over and over again in my head when I was young? God, she saved my life! She drove me to the hospital the morning of the stroke. She did the best that anyone could have expected. She took on the responsibility of being the mother to Mark and Betty Rae during my absence. She cleaned, cooked, worked two jobs, and attended classes at the community college. If she is depressed, how can I help her? I want to wrap my arms around her and tell her everything is going to be all right, but I can't do that right now. I sense that she is afraid of me and I'm afraid, too—afraid of her rejection. She understands the uncertainty we are all going through. It is as if we don't know each other anymore. She acts as if she is angry, fearful, and frustrated at the metamorphosis in her mother. Her grades and her studies are reflecting this dramatic change. She isn't making time for her studies. She is afraid of losing her job. Paige is afraid for her father too. Both of them think of me as another child. It must be frustrating for her. Paige deserves to be a normal teenager. She needs her mother. She seems to resent me or the stroke or both. She feels cheated. She needs time to understand and deal with this frightening experience. She never expected this to happen. It was just too sudden. Mothers don't get sick. She hates my slow movements, monotone speech, and limp. My voice is irritating to her. I can see it in her face. She doesn't see the fairness of this disease, because there isn't any. I know she wants to blame the doctors who should have been able to see and diagnose this problem before it got this far.

She treats me like one of the children in child care when I need help tying my shoes or getting dressed. I know that it can be frustrating, difficult, and demanding, but I need her understanding. I pray that time will heal her wound as well as mine.

Seizures

SEIZURES CAN BE VERY FRIGHTENING, especially when they occur during recovery from a close brush with death such as Cleo has had, but they are not a very common complication of stroke. In the great majority of patients, seizures that occur after strokes are relatively readily controlled with medicines.

The brain is basically an electrical organ, made up of billions of neurons that transmit their messages electrically at synapses. Normally, the electrical activity of these cells is related to activity in the environment and inside the body. A seizure is a kind of inappropriate activity or short circuit in a part of the brain during which some nerve cells discharge spontaneously and hyperactively, sometimes even without a stimulus. The nerve cells that discharge are usually abnormal neurons that have been injured. Excessive discharge of local neurons is sometimes quickly spread through the nervous system and a seizure occurs.

Seizures take many forms. The seizure can be a motor seizure, with relatively violent contractions of the muscles of the limbs and face and jaw. Often the muscles stiffen before they rhythmically jerk. Pelvic and abdominal muscle contractions can cause a release of urine and bowel contents during a seizure. During these seizures the patients are unconscious, and afterward they usually do not remember the attack. In other patients, the seizure spreads to both sides of the brain, and the patient looks blank and momentarily loses consciousness but does not shake. Sometimes the discharge remains local and patients show localized jerking of muscles within the weak side. Sometimes the local discharge causes a visual, tactile, auditory, or olfactory experience or an inappropriate emotional feeling.

When seizures complicate strokes, most often they develop after the acute stroke, when some nerve cells have partially recovered. The initial seizure or seizures can be quite alarming to observers, although patients are most often unaware of what happens during a seizure. Usually seizures are followed by a period of drowsiness and decreased alertness. The nerve cells, having discharged vigorously, have used up a good deal of energy and need time to recover. Headache, aching muscles, and a bitten tongue are also often present after motor seizures and usually reflect increased muscle activity during the seizure. Patients report feeling as if they had been beaten up or as if they had run vigorously. At times, after a seizure, patients can become temporarily violent and strike out at individuals who are trying to help them. On awakening from a seizure, patients often recognize what has happened because of a gap in their recall, aching muscles, a sore, bitten tongue, or evidence of incontinence.

Seizures almost always stop spontaneously. Observers need not put anything in the person's mouth during a seizure—the idea of swallowing the tongue is a myth. The seizing individual should be placed on a soft bed if possible, or at least in a place where the risk of injury is low during the shaking. If it is the person's first seizure, he or she should be taken to a hospital. When someone is known to have seizures, a trip to the hospital after an episode is unnecessary, unless the seizures are repetitive or the patient fails to awaken within ten to fifteen minutes. The treating doctor should be notified later so that medications can be adjusted when necessary.

Illness and the Stress It Places on Close Family and Personal Relationships

CLEO'S 3 A.M. SEIZURE and Larry's shocking response are a wrenching example of the strains that a lengthy illness can place on close family and personal relationships. Cleo goes on to describe not only the devastating effect of her seizure on her husband, but also the considerable evidence that he is having an extremely hard time adjusting to the vast changes the stroke has brought into their life. Beset with economic worries and faced with a wife whom he considers handicapped and who has intermittent incontinence and seizures, Larry is severely stressed by the change in his wife's health. Other adult relatives live too far away or have been unable to come and stay long enough to offer respite in these difficult early weeks after Cleo's homecoming. Doctors often forget that caregivers sometimes need almost as much attention as the patients. Caregivers can become so stressed that they cannot cope. They may neglect the patient or act out in an angry, almost abusive, fashion. Some discussion and counseling can be given to caregivers when patients leave the hospital or rehabilitation area, but as problems arise, doctors must be prepared to continue to work with caregivers and be watchful for caregiver breakdown.

Cleo's older daughter, Paige, also feels the strain of added responsibilities and the lack of parental guidance. She has become a caregiver. The usual dependency role between parent and child has abruptly been reversed and may be difficult to recapture. The strains on the husband-wife and parent-child relationships are enormous. Illnesses happen to families not just individuals. Cleo is clearly correct that the family needs help and advice. At times, the stresses become so great on the caregivers that they choose to selfishly bail out of their relationships and responsibilities. Despite the fact that the marriage vows say "in sickness and in health," illness and time often change a loved one from the individual he or she was to a totally different person. Some significant others and family members simply cannot cope with the change, the stresses, the responsibilities, and the shift in attention from the *me* to the *thou*.

Doctors and nurses should emphasize to caregivers that they may need to pay even more attention to their children after a spouse's stroke. The patient who had the stroke played a large role in the life and development of the children, and after a parent's stroke the children may have as much—or more—

difficulty as the spouse. Similarly, when a child is ill or has a stroke, the care-giving parent must be counseled not to neglect his or her spouse. The family must be kept together, and spirits should be upbeat to meet the challenges. If this counseling has been given to Cleo's family, it hasn't done enough. More and more, as she fights her way back from stroke's wilderness, it seems her family is becoming lost.

A MARRIAGE IN TROUBLE. A sexy silk negligee lay in the dresser drawer with the tags still on

March 1993

it. During our anniversary on December 31 it remained hidden away.

In March, Larry was awarded a "winner's trip" to Florida for in-creased sales over the past year. I have debated about whether I am ready for such a trip only nine months after the stroke. The doctors assure me I am able to go. The negligee is packed and unpacked, but I need Larry to hold me and find me attractive once again.

This evening we are alone in our hotel room. I set the mood—the low lighting, soft music, cool breezes, his favorite perfume, and me. After a few awkward moments he turns to me, "I'm sorry, Cleo. I can't make love to someone who's handicapped." The words bite deep. I retreat to the bathroom to cry.

I have spent much of the first year in one bathroom or another, cry-ing. I have tried desperately to figure out how I appear different. I know I favor my left side and I know my shortcomings. Why does he have to tell me again?

I realize that my husband also has a challenge. His challenge is to accept me. At first, he was thrown into a time where he didn't know if his wife would live. Then, as I recovered, I could not work and he felt the loss of income as well as his companion. The increased debt, selling the house, grocery shopping, cooking, cleaning, and single-parenting were all his tasks, besides his full-time career. Am I cast out? Am I expendable? Am I damaged goods?

We must seek family counseling. Each member of our family is hurting. We have all lost something. It will take time to heal. Time is a gift, not my adversary.

I have been told that mood swings and depression are a part of stroke. It will take months or even years to see what my full capabilities will be. I always thought that I would heal and recover fully, although I was not healed upon hospital discharge.

The doctor's words echo in my ears, "This is permanent." Do I have to go through this alone, without surrounding myself with people who love me and in whose presence I can admit my vulnerability? I know now that I will not allow myself to heal completely until I release the anger within me.

Sex and Stroke

SEXUAL INSTINCTS AND ACTIVITIES are important, lifelong aspects of humans. Sex is just as important and vital in middle age as in youth. Strokes can seriously impair sexual functions and lessen normal sexual activities. The effects of strokes are multiple and include physiological, practical, and psychological factors.

The brain, spinal cord, and the rest of the central nervous system control sexual functions. For a stroke patient, sexual responses and performance can be temporarily lost or diminished, but are usually not permanently affected. The brain has more to do with desire for sex (libido) and for transmitting desire for sex to the genital organs. Libido can be diminished or increased by strokes. In some patients, the stimuli that aroused sexual feelings before the stroke may not cause the same arousal after the stroke, depending on the stroke's location, type, and size.

Psychological factors, and not just those of the patient, are also very important. Although Cleo was clearly interested in resuming normal sexual activity, dreaming of Richard Gere and plotting to seduce her husband, her husband clearly expressed his reluctance to "make love to someone who's handicapped." Larry was psychologically unprepared to return to sex and found his wife less attractive and less desirable. Could his fear that she might lose control of her bowel or bladder function during sex have subconsciously or consciously played a role in his mind? *Was* she "damaged goods"? Did he somehow fear that sex could provoke another stroke or seizure?

Many spouses are afraid that sex will precipitate a heart attack or stroke in their partner. Some caregivers see stroke patients and patients who survive any serious illness as somehow more fragile, breakable beings after recovery. Since sex usually involves vigorous physical activity, caregivers are somehow fearful that the patient will "break" in some unclear way. These caregivers are overprotective in every other way. Other spouses may not have been very interested in sex even before their partner's stroke, and they use the stroke as an excuse to avoid sex. Some concerns and fears of the stroke patient's partner can be allayed by open discussions; other fears are vague and unspecific and can't be easily overcome.

Of course, some patients are also fearful and psychologically unprepared to resume sexual activity. They often fear that their performance will prove wanting to their partner. As a result, this new timidity and self-consciousness can make them seem less a man or less than an attractive woman. Some stroke patients are fearful that sex will precipitate another stroke, heart attack, or seizure or somehow injure them in some other way. Some fears can be allayed by simply trying and accomplishing intercourse. Sometimes, sexual activities other than intercourse can be more easily performed. These should also be encouraged. They help convince the stroke patient that he or she is still desirable and attractive sexually.

There can also be practical barriers to sex. Motor deficits may make positioning more complex and mobility of the trunk and limbs can be reduced. Agility in attaining manual, oral, and genital stimulation may be affected. Loss of sensation in various regions after the stroke may decrease previously normal sexual arousal.

I suggest that the patient and spouse or other sex partner discuss these issues openly with each other. Physicians and therapists who specialize in sexual issues can be consulted. Each partner must be patient with the other. Many patients are able to resume normal or even heightened sexual activity after a stroke.

June 1993 Today a manila envelope came in the mail. It is addressed to Mr. and Mrs. Larry Hutton and has the return address of a legal firm. I vaguely remember Larry and me consulting with a legal firm regarding the original MRI taken in May 1992. Our question was, if that MRI indicated a small stroke, why didn't the attending neurologist do something to prevent the stroke that inevitably changed my life on June 9, 1992? It is a natural question in the process of healing. We needed answers when there were none. Originally, we lashed out, wanting to blame someone for this tragedy. The MRI was sent to an expert radiologist, and the letter states that a very tiny lesion was detected. It goes on to say that we do not have legal recourse, because the care and medical precautions that were afforded me that day met all the standards of medical practice. In retrospect, I realize that I had had a TIA (transient ischemic attack) and, at that time, there was no treatment available to prevent the subsequent stroke. We did not proceed with any legal action.

Enclosed with the lawyer's letter is a small cassette tape marked "Larry Hutton." With the children in school and Larry at work, I retreat to the bathroom, scrunch down on the carpeted floor and play the tape. The tape contains Larry's voice as he tries to express his feelings about the stroke for the lawyers. He talks about how our life has changed after the stroke. But does he know *I* am the stroke? I listen to it over and over again while staring at my image in the mirror. I hate myself. I hate my deadpan, expressionless face. I want to grab through my skull and pull it out of my brain. I want my personality back! I want all the memories of who I was quickly implanted back into every cell and fiber of my being. I don't want to be like this anymore! Oh, God . . . please . . . fix this! Slowly muttering every syllable in the mirror while drool slips down from the corner of the left side of my mouth, I meet the reality of stroke. My family isn't healing, again Larry asserted his true feelings, although probably not meant for my ears. I had heard his words and they hurt like razors. There was no way to hurry the healing process. All I can do is shelter myself from any more emotional pain and try to plug the dam of hurt oozing from my family.

Flipping the tape over to the other side, I hear Paige. My poor girl is going in circles, and the family is spinning out of control. I click the button on the tape recorder and sit in the dark corner of the room, rocking back and forth, trying to absorb it all. I will separate myself from them and somehow salvage what is left of my crummy life and allow my family the right to get on with living. I must somehow stop hating myself for something I had no control over. I have to heal or surely my soul will die.

Personality and Emotional Changes After Stroke

IN THESE VERY PERSONAL and troubling tape recordings, Larry and Paige explain in detail how Cleo's stroke has changed everything. They also emphasize the dramatic personality changes that they recognize in Cleo. She is not the same person. Personality changes and changes in interpersonal behavior are common in stroke patients but have not received as much attention as changes in language, memory, and visual abilities. Personality and behavior change most after strokes that involve the right cerebral hemisphere, the more emotional—which we call "affect-" or mood-related—part of the brain.

Patients who, like Cleo, have right cerebral damage often have trouble showing emotions in their facial expressions and voices and may have difficulty picking up the body language and emotional tones of others. We call this *dysprosody*, which literally means an abnormality of the rhythm and tone of speech. Speech has two major components. One is linguistic and relates strictly to the meaning of the words used. The other is affective: by the tone, volume, accent, facial expression, gestures, and word emphasis, we can give the same words very different meanings. The words "Come home early" when uttered by a spouse may have many meanings depending on their context and how the words are spoken. They could mean "If you get a chance, it would be nice if you were early," or they could mean "You better be early or else there will be recriminations." The words could have a sexual or other connotation indicating some reward for an early appearance. The ability to transmit and interpret body language and emotional tone is an important part of communication between individuals. Strokes, especially those in the right cerebral hemisphere, can blunt affective responses and alter nonlinguistic communications. The brain research that has allowed us to better understand these effects of stroke has been done quite recently—in the last decade or so—but it is convincing, and being able to more clearly account for these problems to patients, caregivers, and significant others may greatly help them cope and adjust. Certainly the source of these problems should be explained to them.

CHAPTER 6

On My Own

June 1993 On June 12, I left Larry and my children. I still have the receipt from the moving company that relocated me to a small one-bedroom apartment about five miles away from our new home. We sold our large home and purchased a more moderate, rambler-style home in the same school district. When we purchased the new house, I had every intention of staying and thought of it as a new beginning. Larry's busy working schedule allowed him little time to look for a new house, so I had found a home for the family. Larry telephoned a realtor and said exactly what type of house we were looking for and I went house hunting with the real estate agent. My body did not cooperate in choosing an appropriate time to use the toilet, but I had chosen a home.

I did my own packing, just a few essential things I would need for a six-month sojourn, and contacted the moving company for a two-step move—one to an apartment and another to a home. I continued to feel like an albatross around my family's neck.

In hindsight, I understand that my thalamus was affected. The thalamus affects flight, fight, panic attacks, eating habits, and sexual desires. I was determined to fight the residual affects of the stroke on my own. We are going to have a meeting of the minds, me and stroke, and I want to fight it out on my own terms and in my own way instead of dragging innocent people into the fray. I am going to find myself, to try to find the person I once was, to tackle what had to be done and give my body and spirit time to heal.

My family support system is nonexistent. The children are busy being children and do not need the additional stress of seeing a ghost of a mother. Larry and I do not communicate anymore. I skip meals and find eating only a chore of necessity. I am alone and unlovable.

I telephoned Larry at work and made an appointment with him for lunch at a restaurant. We agreed on a six-month separation. I said that I was attending counseling sessions and expressed that it might be a good idea for him and the children to go as well. Within a few days, I moved into a small apartment that my social security income could afford. And now, a couple of weeks later, the family has moved to their new house.

The apartment is in an old Holiday Inn that has been converted to rental units. My apartment, Room 109, consists of a long hall—to the right a small bathroom and a bedroom large enough for a single bed; to the left a room with a kitchenette. It is all I can afford. The apartment complex is located in a hollow behind a gas station and convenience store on the frontage road of a major highway. The residents vary from single elderly men to young people with even younger children. Maybe the fear of being on my own and feeling vulnerable makes me slip off my engagement ring and tuck it in a safe place. Maybe it's the rumors that float around about drug deals and police being called to the units in the middle of the night that frighten me.

Now that my outpatient therapy is complete, it is up to me to put the pieces of my life back together. I'm in over my head and I know it. I can't count money, and keeping a correct tab on finances is an incredibly frustrating task. I can't remember if I completed a chore or where I put things. Finding a notebook, I begin to organize my life.

To help defray medical costs, I begin a letter-writing campaign to our medical creditors. My cardiac surgeon has responded, stating that under the circumstances he will not attempt to collect the difference between his fee and what the insurance covered, and we owe nothing more. We otherwise would have been obligated to pay what our insurance plan did not cover.

September 1993 I have registered for classes at the area community college to keep my mind occupied. Although my official therapy sessions have ended, I am not finished with rehabilitation. A bus service picks me up at the door each day and brings me to and from the college campus. I am receiving state aid for my education and remain on Larry's medical insurance coverage for my medications, which are delivered to my door. I slowly learn to tell time again and set an alarm clock in order not to miss the bus for classes. I tape-record the few classes I attend and repeatedly listen to the lectures. My life revolves around going to and coming home from the campus, as I do not drive now. I frequently telephone the children, and once in a while they stop by for a visit. I have all the time I need to learn. I try to erase from my mind things that are out of my control.

October 1993 I have been waiting for this Stroke Conference, held at the Courage Center, for some time now. I register for Larry and me, as he agreed to go with me, and I'm hoping that we'll learn a lot about stroke. After the morning session, Larry appears tense and fidgety. I ask if he wants to stay for the afternoon session or go to my apartment. We decide to go out for lunch and then drive to my place. We begin to kiss but it is mechanical and different. We both realize it is best for him to leave. I surmise he isn't ready to accept so much in one day.

November 1993 On Thanksgiving Day, Larry drives over to the apartment to get me so that I can prepare a turkey dinner for the family at home. I look forward to the challenge, and the family is going to help. As we drive up the street, Larry pulls the car over to the side of the road. "I want a divorce," he says. "You get an attorney and we'll . . ." My mind goes blank. His words pass as I sit in resignation. My body tightens, and I close my eyes and my mind to get through the day. That evening, I fall asleep on the couch, and the next day Mark tells me that Larry covered me with a blanket. As Larry drives me to the apartment, we cannot find any words to share.

December 8, 1993 Within the six months, Larry has tried to get his life together, too. He dresses differently; he has replaced the blue shirts and mottled brown suits with paisley shirts, dark, tailored suits, and loud ties. He looks different, more self-assured, with a brilliant smile due to new dental work. He smells different; not that my senses are any more attuned to new fragrances, but it isn't his usual cologne.

I HAVE BEEN TAKING COUMADIN to prevent further heart problems ever since the heart surgery seventeen months ago. Monthly, I have my blood taken at the hospital laboratory and checked for the clotting time; the medical community refers to this as a *prothrombin*, or *protime* blood test. The doctor tells me whether to increase or decrease the dosage of Coumadin depending on the lab results. I have been taking five milligrams, or one tablet, on Monday and Friday, and two and a half milligrams, or half a tablet, the other days, and I mark a large calendar hanging on the refrigerator after taking the dosage and note how much I have taken.

ONE MORNING I AWAKE slightly groggy and confused. My actions seem slow and deliberate. I have managed to pull on a pair of jeans, but the process of how to slip a sweatshirt over my head suddenly escapes my ability. By midmorning I am able to hang on to the walls and drag myself along the short hallway from the bedroom to the kitchen counter. Picking up the telephone receiver, all the numbers appear jumbled. I press the preprogrammed number of my friend Fran.

"Hello."

"Fran? Fran, I need help."

"Cleo, I'll be right there."

I lay my head down across my folded arms at the table and fall back to sleep again.

"Cleo, it's Fran," she shouts, opening the door with the extra key I had given her. The drapes are drawn and it is dark in the apartment.

"Cleo, what's going on?" she asks, half out of breath. "Talk to me. Come on. Tell me what happened." I hear her, but I can't respond appropriately. Thinking about every word makes me more tired and confused. "I'm calling the doctor," she says.

She informs me that we are going to the hospital. I feel relieved but at the same time angry because I don't want to go through it again—the expense, the backsliding, the time spent building my body and spirit up just to be slapped down.

The emergency room staff admit me, do blood chemistries, and order a brain scan. In the evening, my neurologist arrives.

"I'm glad to see you more awake. The scan did not show any evidence of a new stroke. I know you can't talk right now and that is upsetting to you. You also have extreme weakness of your left side again and your protime is 54.6, which is quite high. It means that your blood is too thin and will not clot for an excessively long time. There may be some micro-hemorrhaging in the brain that would not show up on the CT scan. I insist that you stay in bed. You cannot fall with a protime that high. You could cause internal bleeding and it could be very serious. I'll start you on vitamin K injections to counteract the Coumadin. As soon as possible, we should follow up with a cardiac workup to see if we can dispense with the Coumadin. Instead, an aspirin a day may work for you as an anticoagulant."

The nursing staff wake me with flashlights shining in my eyes and ask me to grasp their hands or tell them what day it is. I am not cooperative. I do not know the day or even the time when I look directly at the clock. Words that I do not consciously choose flow out of my mouth.

The speech and physical therapists begin working with me again. The left side of my face droops and I have trouble swallowing. The speech therapist gives me thick liquid shakes and teaches me to bend my chin to my chest when swallowing.

However, I am tired of being a patient. My neurologist understands that I do not want anyone notified that I am in the special care unit. It is my way of accepting the fact that this will not be a long hospitalization. I do not want to impose any more stress on my children's lives.

The protime levels are still climbing, and I am not taking the Coumadin. Another injection of vitamin K is ordered. Within a few days I am walking with a cane and talking slowly but purposefully. My comprehension and attention span are short. When I attempt to focus my eyes for reading, the words look familiar but I can't remember how to say them. Within five days I am discharged from the hospital.

January 1994 I refill my prescriptions. The pharmacist asks, "You're not taking Coumadin with quinine sulfate, are you?"

"I was taking both for five months. I am not taking Coumadin now. Why?"

"They conflict, that's why! It's just as bad as taking aspirin and Coumadin! The quinine stores the effects of the Coumadin so it accumulates in the bloodstream and makes it very potent over time. It could even be lethal!"

When I return home I telephone my neurologist's office for an appointment. Within a few days I am sitting in his office. I take out the dark-colored plastic container of quinine sulfate and the remaining Coumadin from my purse and place them on his desk. "Do you see anything wrong with taking these two medications together?" I ask.

"But you're not on the Coumadin now, are you? Anyway, it is probably for the best since you should have been off the Coumadin by now. It's over, Cleo, and you're okay. Taking one aspirin a day can control your blood levels and you won't have to go through that monthly blood work. What are you taking the quinine for?"

"The muscles in my left foot and calf are pulling and I am having trouble walking. My general practitioner gave me quinine sulfate last July. He knew the other medications I was on, because he monitors my protime and adjusts the dosage of Coumadin!"

I wanted to place the blame for the unfortunate incident on the doctors or the pharmacist for not catching it earlier—anyone but me. But I realize that everyone has a responsibility in the giving and taking of medications.

Heparin, Coumadin, Aspirin, and Other Drugs Used to Treat Stroke Patients

A COMMON STRATEGY used by doctors to prevent stroke recurrence is to modify the clotting tendency (called coagulation) of the blood. The aim is to prevent clots from forming in regions where blood vessels are narrow or irregular. Brain ischemia is caused predominantly by vascular occlusion, and blood clotting plays a critical role in blocking the vessels. Of course, if the stroke is due to hemorrhage, giving a medication that will reduce the ability of the blood to coagulate is potentially disastrous.

As I mentioned in Chapter 4, two different types of agents are used to alter blood coagulability. One type mostly changes the stickiness of blood platelets. When a vessel becomes irregular and a plaque forms, the altered inner lining of the blood vessel stimulates platelets in the blood on the surface of the plaque to clump together and stick to the wall of the vessel. This creates a small white clot that is made of platelets and another blood substance, fibrin. The white clot can later be displaced from the vessel and can be carried to vessels within the brain, causing transient ischemic attacks (TIAs) or strokes. To discourage platelets from clumping together and sticking to the vessel walls, drugs called antiplatelet agents are given. Aspirin, ticlopidine (Ticlid), clopidogrel (Plavix), and dipyridamole (Aggrenox, Persantine) are the best known such drugs. Most of the nonsteroidal drugs used for arthritis and pain relief also have antiplatelet actions; these drugs include ibuprofen and indomethacin. Some natural food substances, such as omega-3 oils (so-called fish oils) and the black fungus commonly used in Chinese food, are rich in substances that affect platelet functions.

Another strategy to prevent another stroke from occurring is to give drugs, usually called anticoagulants, that decrease the tendency for red clots to form. A natural substance, heparin, or similar substances are often given first by injection or intravenously while stroke patients are in the hospital. Later other drugs, including Coumadin, are given by mouth. Coumadin decreases the clotting tendency of the blood and makes spontaneous bleeding and bleeding after a cut or injury a potential problem.

To monitor the tendency for clotting, doctors order blood tests to compare the patient's tendency to clot against normal standards. The tests are called prothrombin time determinations, and the results are given in time (for example, 14 seconds) or in a ratio of the patient's prothrombin time to local controls in that lab (say, 2 times the control) or as a ratio determined by comparison with an international standard, the INR (international normalized ratio), leading to readings such as an INR of 2.1. The blood must be monitored frequently at first to make sure that the results show enough reduction of clotting tendency without excess risk of bleeding. Usually the doctor or his or her nurse or physician's assistant calls the patient after receiving the results of the test and tells the patient how many pills to take that day and subsequent days and when to have the next blood test.

Blood clotting tendency and the effect of Coumadin are influenced by many other drugs and probably by some foods. Coumadin dose is also affected by hormonal changes in women associated with menstruation, pregnancy, ovulation, and use of oral contraceptives and female hormones. Some of the factors affecting the dose of Coumadin and its effects are not known.

Cleo's prothrombin—or clotting—time (50 seconds) was much too long. Doctors try to keep the prothrombin time at about 18–19 seconds or slightly lower. This usually equals an INR of between 2 and 3. Her very high level made her a risk for bleeding. I don't understand how the high prothrombin time was related to her symptoms, because significant bleeding is readily detected by CT, and her CT did not show a hemorrhage.

Coumadin can be an effective medicine and yet is a risky drug to take. Patients and their doctors must be fully aware of the intricacies of its use, of drugs that alter Coumadin dose and effect, and of early signs of internal bleeding.

CHAPTER 7

Gaining More than the Stroke Had Taken

I BEGAN COLLEGE in the fall of 1993, fifteen months after the stroke. My original goal was to fill time with something constructive. I did not think of graduation. The future seemed too far away. Instead, I focused on the present. All I knew was that I needed more rehabilitation. After outpatient therapy, it was up to me to get back some sort of normal routine. But everything seemed so far from normal that I didn't know where to begin. Although I was very fortunate to have had the opportunity of long-term therapy and encouraging staff that told me, "In time, you will improve," I wanted to recover now! I felt as if I was in a race against time. I had to catch up on all the time I had spent in the hospital, in rehabilitation, in recovery; I had to "hurry and get well quick" like all the well-wishing pronounced by the get-well cards.

I could not return to my job as an in-home child care provider, and my personal life was beginning to fray under the strain of financial burdens and emotional issues associated with the divorce and impairments caused by the stroke. I needed something positive; a sense of purpose. I wanted people around me who wouldn't continually ask me how I was doing, because quite frankly I didn't know. Finally, I determined that a learning environment filled with activity was not only necessary toward recovery but compulsory for my emotional health as well.

At the local community college, I was required to take college entrance tests to see if I was a worthy candidate. With a sharpened number two pencil, I was ready to take on the world. However, my brain was not. I couldn't comprehend the questions or the directions. Logic questions demanded more memory skills. Each word within a question sounded familiar, but by the time I could make meaningful connections, the time was up and the test was over. This was impossible! I had taken on too much too soon. I was unable to finish the test and went home defeated.

As timeless lines from *Gone with the Wind* reverberated in my head about Scarlett O'Hara and her determination to rebuild Tara, I was just as determined to rebuild my brain. I would have to tell the college admissions counselor that I had had a stroke. The quintessential "note from my doctor" provided me with access to Disability Services, additional time for testing, and audiotapes so I could hear the questions as well as visualize the words. I took the entrance exam again. The results indicated that I had barely reached the normal range and math and comprehension skills indicated below-average scores. But I had passed! Academia had allowed me to enter its hallowed halls.

I'd start out slow. After all, I was fully aware that my brain didn't process information on anything but low-speed setting. I took a course in Competency Based Education. Maybe I could convert the education courses I had taken before the stroke and design my own degree. Maybe I could take a shortcut! Actually, Competency Based Education taught me how to define my educational goals, gave me choices of different educational programs, and taught me what it would take to attain them. It wasn't going to be easy. It wasn't going to be quick.

With the goal of an Associate of Arts degree in mind, the next semester I took two classes, one in Piano and the other in English Literature. It may sound like nonsense to take a course in Piano when I had left-sided weakness, but before the stroke I was quite adept at hammering out a mean "Chopsticks." Besides, I thought it would be good therapy for fingers that didn't want to move on command. I practiced before and after class. I practiced alone in solitary rooms where I could swear at the awkwardness of my fingers and cry because I couldn't understand where the notes belonged on the keyboard. But not in class! In class, I was a student, not a crybaby.

In class, I asked professors for permission to tape-record their lectures, as this helped memory retention. I tried taking notes, but lectures seemed too fast for my pen. Disability Services provided quiet environments and a little more time for testing. Independent Studies assisted me through winter months when ice conditions threatened my unsteady steps. But I enjoyed being in the classroom, being with other students, walking through the cafeteria line and trying to balance a tray with one hand, carrying books in a backpack, walking hallways, attempting stairs, conversing with friends, studying, and being a part of something bigger than myself.

OF COURSE there were failures too. One term I tried to take math and a foreign language course in order to meet college requirements. I studied for hours trying to do simple mathematical computations only to find a big red F on my paper the next day. The foreign language class was just as difficult, as I was just learning to speak and comprehend English again. I dropped the foreign language class and decided that I would attend to one problem at a time. Even with a tutor, my math skills weren't improving. Eventually, I consulted with my neurologist and asked why I couldn't perform simple addition or subtraction problems. He wrote two double-digit numbers on a piece of paper and asked me to add them. I tried adding the numbers to the left as I had learned in elementary school, but where to place the numbers or how to write the numbers was erased from my memory—although I was not aware it was a stroke deficit. Instead, I had put hours of study into something beyond my capacity at the time. I could not understand why I could not add ridiculously simple double-digit numbers. The math course was much more difficult than the mathematical problem posed by my neurologist. With a letter from my neurologist, I substituted logic for math. Even though, at the time, reaching logical conclusions wasn't my forte either. I passed the course.

Memory and the Organization of Cerebral Hemisphere Functions

AS FAR BACK AS her first days in the hospital, Cleo has repeatedly mentioned difficulty remembering things. The structures thought to be most important in making and storing new memories are shown on the facing page.

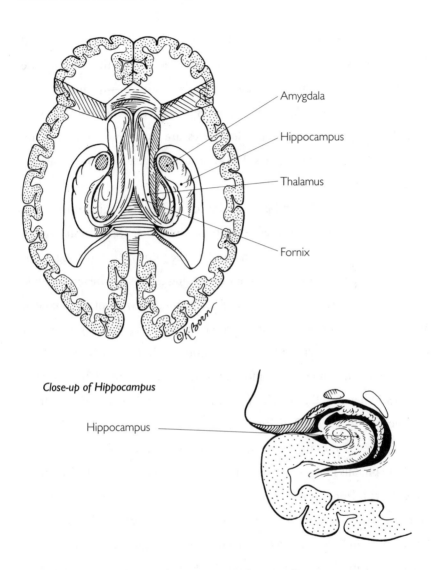

Amygdala

Hippocampus

Thalamus

Fornix

Close-up of Hippocampus

Hippocampus

Structures thought most important for memory. *The hippocampi and the amygdaloid nuclei play important roles in the retention of memories. The hippocampi, from the Greek for "sea horses," whose shape they resemble, are deep in the brain near the temporal lobe part of the lateral ventricles. When both hippocampi are damaged, a person cannot form new memories. The amygdaloid nuclei lie next to the hippocampi and are almond shaped. They seem to be important in forming memories with emotional content. These structures connect to the thalamus by a band of tissue called the fornix. Some nerve cells within the thalamus also play an important part in memory.*

Memory storage is located mostly in the middle parts of the temporal lobes on both sides of the brain. Two structures located in this area that are greatly involved in memory processing and the emotions—the hippocampi and the amygdaloid nuclei—play important roles in the retention of memories. The hippocampi, shaped like sea horses, are located near the temporal lobe part of the lateral ventricles. The amygdaloid nuclei lie next to the hippocampi and are almond shaped.

Memory functions can be divided into three parts: registration, reinforcement and storage, and retrieval. To retain information, an individual must be attentive and interested in recalling the information and register it in the brain. If you are thinking about something else, or daydreaming, while being told something, you will not register it and thus will not recall it later. You can retain information better if you keep repeating it, or if you relate it to something else to reinforce it. Information is probably stored in files, like in a filing cabinet. You retrieve information using your frontal lobes to search the files. When asked to recall their third-grade teacher, most individuals somehow try to enter that file by imagining themselves in the school that they then attended. They might recall their mother sending them lunch, or the little girl in pigtails in the row next to them, or the teacher in front of the class. Patients with frontal-lobe strokes often have difficulty retrieving memories and organizing them according to time. Patients with temporal lobe lesions, like Cleo, often have difficulty storing memories. The left temporal lobe and thalamus are more specialized for word and language memories, and the right temporal lobe and thalamus for visual memories.

May, 1995– In 1995, I graduated with an Associate of Arts degree from community college. It was a good year. I had lived independently for two years, and our son graduated from high school the same month I graduated from college.

It was time to make another choice. I could either move back to my hometown and complete a Bachelor of Arts degree or stay in the area close to our children, who were 21, 18, and 16. Our two oldest children assisted me in moving to my hometown during one of the worst snowstorms in our state's history, February 2, 1996. I could either revel in their dedication to their mother's higher learning or consider the option that my children relished their freedom. I decided that it was a little of both as I recalled Oliver Hardy and thought, here's another fine mess I've gotten myself into. In the spring I transferred to the university campus in my hometown.

With the transfer came a resurrection of old academic records from my earlier attendance at that university, circa 1968. I would have to contend not only with stroke deficits but with a lowered grade point average due to my younger, "flower child" days. Again, like Scarlett O'Hara, "I'll think about it tomorrow" became my constant mantra. With the help of the Department of Rehabilitation Services and the disability Access Services offered through the university, I eventually weaned myself from students who had volunteered to take notes for me and Books on Tape. The educational process that had begun as a coping mechanism had developed into a very positive goal. While at the university, I regained a keen sense of humor—one that demonstrated with a smile, "Quiet, Brain Work in Progress." I studied before classes began in order to store information in long-term memory. I worked so hard listening, trying to take notes, and following the professor's lecture that unconsciously my affected left arm would spastically wave in the air, signaling the instructor that I had a question, but when called upon I wouldn't know what to say. I often sat on my left hand or clutched it between crossed legs so I would know where it was at all times. I studied for exams, only to forget which classroom they were held in. I kept a calendar and notebook of classes and appointments that would have rivalled that of the president of the United States. I practically lived in the library, surrounded by common words I could not always understand. I learned to use a dictionary when I became confused. In my spare time, I'd make copies of several cities' daily newspaper crossword puzzles and try to complete them.

In 1999, I graduated from the university with a Bachelor of Arts degree in English and a minor in Professional Writing and Communications. I had retaken courses from the sixties in order to reflect my present academic standing. I had attained the Dean's List in the College of Liberal Arts on several occasions and graduated with Distinction in English with a 3.3 GPA.

The diploma from my state's university proclaims "The Regents of the University . . . on recommendation of the faculty have conferred upon Cleo Hutton the degree of Bachelor of Arts with all its privileges and obligations given . . . this twenty-first day of May, nineteen hundred and ninety-nine." Like the scarecrow in *The Wizard of Oz*, I realize that a piece of paper does not truly reflect all that I am. It is merely a measure of intelligence, commonly used as a stepping stone to a career. The act of attending classes and getting good grades did not resolve all of my stroke deficits. However, the classroom was a powerful tool that assisted me through the rough spots of recovery. I had staved off depression by using humor intertwined with my favorite old movie lines and themes. I had been as tenacious as James Cagney: "You're not going to get me, you dirty rat . . . I'm going to beat this rap . . . ya hear!" Humor, a positive attitude, good study habits, friends, and available services had pulled me through.

The diploma did not bring fame, glory, or riches. But the years invested in education continually exercised my brain until I eventually gained more than the stroke had taken; I had "mastered the tasks," "stayed the course," and "lived up to my potential." The diploma does not state that at the time of graduation I was a 50-year-old stroke survivor, but I knew and now you do too! The diploma gave me courage to conquer other barriers as well.

After graduation, I took a driving course offered by the Courage Center, and I'm on the road again. After graduation, I began giving presentations on stroke awareness and recovery issues. I continue to advocate for stroke survivors and educate the public on issues about stroke recovery. Two ischemic strokes will always be part of my history, but it is humbling to see the resilience of the mind and spirit. As the pharaoh (Yul Brynner) said in the movie *The Ten Commandments*, "So let it be written, so let it be done."

July 2002 My 4-year-old granddaughter, Paige's daughter, and I decided to go on a walk to pick wildflowers for her mommy. We found some pretty purple ones and added them to our bouquet of dandelions and cowslips. We walked back home to investigate the side of my house, where lilies of the valley and forget-me-nots bloom. I explained to her that the name of the little blue flower is forget-me-not, and she said, "Look, Mommy! Look at all of grandma's don't-forget-me's!"

I THINK, DEAR READER, this is an appropriate way to end my journal of recovery. God has a way of planting seeds; the gift of growth is up to us.

Afterword

IF THERE WAS A WAY I could have avoided the events of the summer of 1992 and the permanent effects the stroke has left on my body, both physically and emotionally, I surely would have. However, I have had the opportunity to look inside myself and learn not to be afraid of my imperfections. I have been humbled by the experience.

For me, the word *healing* means to smile, then to laugh. Only then am I able to make the best of the situation. Depression is too familiar to me. There were times I could not get out of bed. I would not answer the phone or talk to anyone. All I could do was watch television. When I knew the television's daily viewing schedule, I also realized I must have been getting better, because I could remember which program came on next!

The pain was intense and I took over-the-counter pain medication like children take candy. I finally realized that the more I was up, moving about or exercising, the less pain I would have and the less medication I would need.

There are times I may backslide. Recovering from stroke is a full-time job and the most difficult thing I have ever had to do. However, the "Why me?" has subsided.

There are people who help stroke survivors: doctors, nurses, counselors, social workers, stroke support groups, clergy, rehabilitation units, friends, and family—just to name a few. I have learned to depend on them for support when I need it.

Over the years I have grown to dislike the word *disability*. By definition it means "not able." It is a self-defeating term. I do not like labels. I dislike politically correct nouns and verbiage that attempts to describe who I am. I am, quite simply, me.

In this Afterword, I will express what, to me, was one of the most difficult parts of stroke recovery, and then I will offer a few other thoughts, born of hard experience, for people who have had strokes and for their families.

The Financial Aspect of Stroke

NO MATTER WHAT happens, physically or emotionally, the financial toll must be paid. A section dedicated to financial recovery is essential. Stroke will be a shock to the family's finances in the best of circumstances, and even worse if you face it single, as I did. When it strikes while there are children to be provided for, it can force a family into a financial corner. Medical and pharmaceutical costs, as well as the costs of vision and dental care, continue to be problems I wrestle with every day.

In 1992, when I had the stroke, I was covered under our family's medical plan through Larry's employment. Prescription costs averaged about $800 per month, and although our private insurance company provided for most of these costs, I knew that with the impending divorce I would have to rely solely on Medicare, which does not cover this expense. I remember speaking to my neurologist about this concern. His response was, "If you don't get the medications, you won't have to worry about paying for a roof over your head!" But when I became dependent on Medicare, he began supplying me with sample packets of the antiseizure medication Depakote. Other ongoing prescription drugs remained financially difficult to obtain.

As the divorce proceeded, medical insurance became a large issue. I would be dropped from Larry's policy when the divorce became final, and the insurance company was already threatening to discontinue my coverage because my medical expenses had reached the maximum, million-dollar, limit allowable. I was not entitled to state medical programs because I was receiving Medicare. I was not entitled to Medicaid because I was not 65 years old.

After our divorce was final, subsequent medical expenses were covered under Medicare and hospital programs designed to assist financially insolvent patients.

I had completed my two-year degree at the community college receiving state and governmental educational aid. At that time, I was not aware of the Department of Vocational/Rehabilitation Services.

After reviewing the possibilities of living in some low-income housing projects in the metropolitan area, each having at least a two-year waiting list for a disability vacancy, I was resolved more than ever not to move there. If I allowed poverty to mandate my state of affairs, I thought I'd never be able to rise above it. My nemesis was not only stroke deficits, but monetary deficits as well. I felt cornered and angry at the world, and no bumper sticker slogan was going to make me feel better.

I had to scramble because my living expenses far outweighed my income. I decided that my goal of getting back to some type of financial stability included a safe place to live within my limited budget, and I began looking for a home back in my hometown. I could complete my education at the university there and be in a location where it would be easier to get from place to place. My uncle found an extremely small home for sale within walking distance of the university and a grocery store and near a bus line. A client of my former child care business was a mortgage loan officer and helped me with "creative financing." Scraping every nickel together, I was able to come up with the down payment and purchase the house a few months short of the fourth anniversary of the stroke. The monthly house payment equaled most of what I was receiving from Social Security. I applied for heat assistance for the winter months and was accepted into the program. My new neurologist and internist, like their predecessors, gave me samples of Depakote and helped me with low-income programs through pharmaceutical companies. Thankfully, I was not taking Coumadin and needed only an aspirin a day as an anticoagulant.

In my hometown, I began working with a counselor at the Department of Vocational/Rehabilitation Services and outlined a plan of action toward completing college.

I learned of the Telephone Assistance Plan (TAP) credit. Operated by Qwest and possibly other telephone companies, this plan offers free directory assistance and a small credit on monthly service after you complete an application and meet disability and income guidelines. The application is reviewed annually.

In an effort to keep property tax within the limits of my income, I applied for disability homestead credit through the state. This is another annual application program that requires income and disability certification.

After college graduation, I began a small press to publish a newsletter for people who have had strokes and for their families. The Department of Vocational/Rehabilitation Services assisted me with a small start-up grant. I wrote a business plan and, with a loan from the local Community Development Corporation, continue to develop my dreams. The business does not offer a king's ransom, but it supplies enough income so that I am considered employed under the state Disability Insurance Program, which currently provides for my prescription needs.

CERTAINLY, I DO NOT recommend divorce. But you don't have to get divorced to have your family finances thrashed by stroke. An overwhelming responsibility comes when the medical bills start arriving, and there will be numerous bills, from every aspect of hospital care and rehabilitation. It is difficult for me to be specific, because every insurance plan is different. Most major medical providers bill the insurance company directly. If a provider does not, it is your responsibility to submit the bill to the insurance company promptly. An idea I used was to keep every provider's bills in a filing system under the provider's name. A person who has had a stroke may have used more than one radiology service, physician, or hospital. It may be easier when the bills are arranged by dates and providers. The best advice I can give is, Don't get worried. Stress is not going to help the survivor or the caregiver. The financial aspect of stroke must be faced and addressed. It may be necessary to consult with your mortgage company, bank, and utility companies and to seek programs offered by your city, county, or state. Above all, don't fret over the big picture but instead accomplish one task at a time. Your medical bills should not go to a collection agency if you contact your medical provider by mail, let them know you are working toward a resolution, and attempt to pay whatever amount possible each month. Don't be afraid to ask for advice from friends or extended family members. If you are a caregiver, include the person who has had the stroke in the decision-making process and write a plan of action for your family. Soon you will realize that you can accomplish each task.

Tap into local community and city programs. Keep some type of file system so that the programs are centrally located and marked in a file in one place. Your community may have door-to-door van or bus service available at a nominal cost if you qualify. When you become overwhelmed with information, reach out to your community health organization for a social service representative who may come to your home and help you complete the necessary paperwork for specific programs. Try not to be demanding but be assertive. Mark a large calendar with each goal and accomplishment.

Find out about state programs available to you because of your age, income, or disability. You will be amazed at the number of ways your state may try to improve the quality of life for people with disabilities. In my state, for example, a disabled resident can obtain a permanent fishing license free of charge. The Senior Federation sponsors Senior Partner's Care Medical Program, which offers health care to those who qualify from providers enrolled in this program. For a nominal annual fee, if you're 65 or disabled, you can use their providers for vision, dental, chiropractic, medical, and legal matters for a discounted fee.

Disability parking tags and licenses are handled by your physician and the Department of Motor Vehicles in your state.

I found resources out of sheer desperation and a tenacious spirit. Each state varies according to the programs offered. Every city may vary depending on population size.

It is also extremely important to remember to bring small pleasures back into your life. If you are able, go for a walk or join your local community center. Invite others to your home for a game of cards. Find a hobby such as gardening, houseplants, fishing, or painting. Go grocery shopping and to the mall just to window-shop and to get out of the house for a while. Attend free community events. Take a course through community education classes or the university for seniors. The Courage Center offers wonderful recreational activities. Bowling is even accessible to disabled individuals.

If you have a dog or cat, remember they need care too! Animals are great emotional healers—the idea here is to begin to care for something outside the spectrum of your own health concerns.

Treat the search for help, both financial and recreational, as part of recovery. The more you do for yourself, the more you will become able to do. Remember, you are your own best advocate.

From "You" to "Patient" and Back Again

I'LL BEGIN THIS LAST SECTION just as I started this book, with the symptoms that turned out to be the forerunners of stroke. Stroke was the furthest thing from my mind. In fact, the diagnosis hit me like a bolt of lightning. It wasn't *my* MRIs the doctors were talking about! I thought that the medical personnel had somehow mixed up the X rays. I felt as if it were a bad dream. This was all in my head, right? Yes, it was literally all in my head.

I had trouble relating symptoms to the neurologist because the symptoms lasted only a few minutes. How could a numb arm relate to a headache? How could tiredness relate to being unable to wear your shoe? The symptoms also progressed over a period of time. How could something I had a few days or several months ago affect me now? I didn't hurt and I didn't have a fever. Suddenly, I just couldn't talk, move, or think clearly. My brain was relating warning signs, but I was too involved in other things to listen.

Please recognize that even the slightest stroke symptoms can be potentially lethal.

Before my stroke, I had many of the warning signs. I knew something was wrong with my health but never considered a stroke. Being unaware of the symptoms of stroke, I couldn't relate long- or short-term specific problems because my mind was not functioning correctly and this caused the physicians difficulty in detecting the stroke. I was having what are known as TIAs.

Transient ischemic attacks, or TIAs, are brief episodes of strokelike symptoms. Knowing the warning signs that can occur just before can prevent some strokes.

1. **Blurred, decreased, or double vision, particularly in one eye**
2. **Difficulty speaking or understanding**
3. **Numbness or weakness in an extremity or face, on one side or both**
4. **Changes in the level of consciousness, unsteadiness, or dizziness**
5. **Sudden personality changes or changes in mental ability**
6. **Difficulty swallowing**
7. **Sudden development of ringing in the ears or decreased hearing**

Do not ignore neurological symptoms. Call 911 immediately.

Family members or coworkers may be very helpful in noticing changes or other symptoms that could warrant immediate attention. Seek a neurologist's examination and opinion immediately. Strokes are hereditary. Make sure your doctor is aware of your family medical history.

To prevent stroke, be aware of your blood pressure, weight, and cholesterol level. Exercise regularly, reduce alcohol intake, and decrease sodium. Have regular checkups with your physician. Sometimes an aspirin a day will be ordered as a blood thinner. However, do not take aspirin if you are taking Coumadin. Check with a physician before taking any medications. At other times, a special diet may be needed along with a physical fitness program.

If you notice any of the warning signs in yourself or a loved one, don't hesitate to seek medical help immediately. If you have increased blood pressure along with hereditary factors of stroke, the message is even more imperative. Do what you can to prevent stroke from happening in your life.

Perhaps you've already had a stroke and this information is hindsight now. When you were in the hospital and therapy, you were referred to as a patient. You were out of your element. You may have felt completely dependent on the staff for food, medicine, bathing, and daily care. Now you are a survivor. The doctors and medical staff helped you, but you are the one who worked both mentally and physically toward your recovery. You may find that you tire easily, especially during the first year. Rest, give yourself time.

The inner workings of the brain continually amaze me. Speech difficulties, immobility, visual disturbances, auditory problems, and pain have nothing to do with a particular body part's capabilities. The brain controls all of our voluntary and involuntary movements, as well as thought processes.

I had difficulty coming to terms with the fact that the doctors couldn't fix the stroke. When a lack of blood flow persists in any part of the brain, some of that part of the brain may die. Once a stroke occurs, there is no way to undo this damage. However, in many cases, a person can be reeducated or rehabilitated. Doctors can prescribe therapy and medication. In some cases surgery can be performed to prevent further damage. It is possible to retrain our bodies and minds to use a new part of the brain.

The main goal of rehabilitation was to recover to my fullest possible potential. In the rehabilitation hospital I was assigned a speech therapist, a physical therapist, an occupational therapist, a recreational therapist, a social worker, and a dietitian, as well as neurologists, cardiologists, and other specialists. Under the circumstances, I could not comprehend this team approach, and it seemed as if too many people were bombarding me during a particularly fragile time in my life. In retrospect, it's clear that each person on the rehabilitation team fulfills a very important function. Even the nurse who read the scribbled notes in my garbage can was only doing it to describe to the doctor my ability or lack of ability in writing a complete thought. The nurse with the flashlight shining in my eyes needed to know my level of consciousness. All this was done in preparation for my eventual hospital discharge and home. A constant assessment of my progress was going on, and I wasn't even aware of it. Both formal and informal tests were conducted to see what my abilities were in the areas of language, speech, fine motor dexterity, visual field, gross motor skills, and emotional and social development. Each therapist performed his or her task very professionally during my recovery process.

I remember the speech pathologist starting out at my bedside when I couldn't utter a single word except guttural sounds. This person also dictated what my diet would consist of, since I would choke on thin liquids. They had to thicken the fluids. The bed was elevated. To prevent choking when swallowing, I tried to remember to tuck my chin toward my chest. Usually the swallowing ability comes back on its own with time.

Because of strokes, on two occasions I lost my ability to speak, and speech pathologists helped me by using a somewhat unusual technique. We started chanting to a beat or rhythm, "Good morning to you. Good morning to you." When I began this practice, it sounded nothing like it should have. But by keeping the beat with my finger or hand, I finally mastered the art of talking again. I believe our bodies work in a rhythm. A stroke is a shock to our bodies, and we must first rest, protect ourselves, and then find that rhythm or beat again.

Music is a wonderful way to relax. In retrospect, my first experience with self-taught biofeedback was listening to music while in the hospital. I have learned to perfect the skill over the years to combat panic attacks.

Speech therapy taught me ways to remember things and how to speak again. We worked with individual flash cards with specific vowel sounds and began working on phonics. I practiced and practiced but could not utter the correct sound. By the end of my hospital stay, speech therapists would read a short paragraph to me and I could answer their questions about what had been read in my slow, monotone, gravelly voice. This exercise increased short-term memory as well as speech.

The physical therapists began with leg exercises at bedside, first by bending and flexing my leg when I was unable to move it. Soon we worked on standing, then walking with a walker, a four-prong cane, and finally walking alone. We practiced going up and down stairs and carefully placing one foot over the other.

Two sessions daily, 30–45 minutes each session, we worked to regain the use of the damaged brain. These exercises must be maintained in the hospital and at home on the affected side, especially for the extremities.

The speech and physical therapy continued on an outpatient basis until the neurologist, therapist, and I felt confident I could continue my own exercise routine at home. Then, I made up a home exercise regime. I wiggled, stretched, and worked with the affected side in every way possible. I would carry a paper or plastic cup in my affected hand and take one, two, or more steps. I walked to and from the college for over two years. Every day, there is something new to do to keep my body and mind active.

Occupational therapy focused on arm movements, dressing skills, and independent living skills. Household chores such as folding the laundry and cooking were relearned. Therapists taught me how to move my head to make the best use of my vision. Because my sense of touch was affected, touching items of different texture became a daily routine and by frequently using this technique I sent new messages to the brain.

I learned that clothing should be comfortable and easy to put on. Shoes must be supportive with a good gripping tread. Shoelaces are available that can be pulled to self-seal. Velcro-seal shoes are also available. Tunic-style tops that fit over the head or shirts with big buttons are best. Men's shirts that have buttons on the left are easier for survivors with left-sided weakness. Pants with elastic waistbands or simple front zippers and Velcro fasteners work well.

When I was discharged from the hospital I kept notes everywhere. The kitchen cupboard looked as though it was wallpapered with notes. It was as if my mind was on the outside. I could not use my mental filing cabinet. Cognitive skills such as understanding speech, money, numbers, sentence structure, and how to interpret body language and verbal nuances were things that had to be relearned.

A dietitian helped me plan a heart-healthy low-fat menu. I had thought that I would never be accepted in restaurants again because I had trouble feeding myself. This was just my paranoia. Now when I go to a restaurant, I order something easy and nutritious. If I need the food to be cut into smaller pieces, I ask the waiter or waitress when ordering. I have learned how to go through a buffet line by balancing the tray with my unaffected hand.

Even today, I may choke on food while dining with another person. My mind works best one task at a time: bite, chew, swallow, breathe, listen, talk. It is no longer an automatic response, but a learned skill.

A person who has had a stroke may have a tendency to eat too fast. It is important to remember that mealtime should be a quiet time, without other stimulation or distractions.

Safety issues continue to be vitally important. You may have to be extremely careful when using a sharp knife, hot water, and the stove because of the lack of sensation or sensory problems on the affected side. I adapted by using the food processor to cut food. I have relearned how to safely use the stove and microwave oven. I test the water with my unaffected side first.

It is a good idea to carry a medical card explaining the medications you are currently taking, your condition, and your doctor's name. You can obtain this card from a stroke support group or the American Heart Association or American Stroke Association.

Today, even though I consistently make sure my doctor knows the exact dosage of every medication I am on, I check with the pharmacist. I go to the same doctor and same pharmacist whenever possible. I ask about, read, or write down the side effects of each medication.

The first three to four weeks after the stroke you may see the most significant improvement. After this time, you may adapt to your challenges by using your own method of accomplishing tasks. Depending on the severity of their stroke, many recovering people will become very creative. In time, you will learn to develop your own style. Therapy will help, but many things will be learned at home. Some people may have to rely on housecleaning assistance. There are social service agencies that can help you with this.

When you are discharged from the hospital you may feel much better. However, you may also find it difficult to relate to people around you. You may become overwhelmed. You may tire easily, become confused, or have little concentration. You may have difficulty with your balance. Noise may become amplified to you or you may have visual difficulties. These may lessen with time. Remember, you are better than yesterday, and hopefully tomorrow you will be better than today.

Humor is important. At some time you will need to laugh. You will have to get your mind off your illness and enter life again. At first you may feel that you may never get there. Depression is a large part of the disease. Mental processing of jokes may be difficult. Repeating them may be impossible. Inappropriate laughter or crying may occur. Maybe a comic movie or a funny family incident will help. I believe laughter is truly the best medicine. It can even distract me from my discomfort. Find your sense of humor. Hang on to it!

Every stroke survivor has to change his or her life, and change is difficult. Time, patience, and adaptation are important factors in stroke recovery. Find inner strength in the face of adversity. Learn to accept the stroke as part of who you are and go on with living. We live in stressful times. Some things can wait until tomorrow. Slow down.

I became more forward with my wants and desires. I am no longer complacent and feel more confident. I want more out of life. Life itself is a new miracle. I find that watching everyone race from one thing to another, buying, accumulating possessions, is just plain crazy. I enjoy the sunsets, the birds, and beauty. I know exactly what I have to do the rest of my life. Educate, contribute, write, speak with others who have had a stroke and their families and caregivers, and enjoy living. Maybe the stroke has taught me that.

A Word to Families and Caregivers

I HOPE YOU UNDERSTAND how my family and I struggled. Because of this, I offer a few ideas to help guard against similar circumstances for the family, spouse, significant other, or anyone who has been touched by stroke. You are a very important part of the recovery process. I desire to give you some insight, alleviate some of your fears, and help you adapt to the many changes and challenges ahead.

A stroke affecting the left side of the body occurs on the right side of the brain, and a stroke that affects the right side of the body takes place on the left side of the brain. A stroke affects every individual differently, depending on the location and size of the damaged area of the brain. Physical disability from stroke is caused when part of the body disassociates from the brain. Aphasia is caused by an impairment of the language center of our brain; it is the inability to comprehend the written or spoken word and, in some cases, the inability to speak. There may be language and communication challenges as the person tries to regain verbal skills. However strange and frightening this struggle seems, the family must understand that their loved one has a working and functioning mind.

Do not feel uncomfortable when visiting your loved one while he or she is in the hospital. Every stroke patient needs support, comforting, and friendship. The family also needs to know others will be there for them, too. Broaden a support system to include stroke support groups, friends, extended family, neighbors, and coworkers.

A helpful hint is to provide the patient with a tape recorder. Place a bold-faced note on the recorder asking doctors and visitors to leave messages for the patient and family. I found this to be a wonderful time-saver for my family and the physicians. It also kept the lines of communication open and clear. For me, this was helpful for memory skills as well. If a tape recorder is not possible, use a large notebook.

People who have had a stroke may have trouble with concentration and comprehension. The stimulation of a television, radio, or too many visitors at once can be overwhelming. Keep the surroundings quiet and restful.

During the acute stage, speak directly to your loved one in your normal tone. Keep your question and comments simple, and wait for the answer. Be observant regarding the survivor's desires. He or she may attempt to communicate in hand motions or spoken sounds. Ask simple "yes" or "no" questions that the patient can answer by a hand grasp or other movement. The patient may use a letter board or may not be able to answer at all. If he or she cannot answer, do not make up an answer. This leads to frustration. Reassure the patient that you understand that he or she can't speak because of the stroke but that with time and therapy this function may return. Avoid giving false hopes. Do not speak to the survivor as if he or she is a child. The patient may have an auditory problem, in which case the message center in the brain is affected. Ask the rehabilitation team for the best approach in communicating with your loved one.

It is very important that the entire family become involved in the healing process because the entire family is affected by the stroke. Reach out early to your support system of counselors, clergy, and doctors. There will come times when frustration, fear, and challenges will become overwhelming. Realize that this is natural and deal openly with the issues. The family needs reassurance and help in accepting this life change. Don't be afraid to ask questions of the doctor or medical staff. Learn about what the survivor's limitations are likely to be, but also realize the potential for greater growth within the spirit of your loved one.

If you have young children in your family, they will react differently to this situation. They may become withdrawn, frightened, angry, or quiet, or they may regress or change sleeping patterns. They need to feel involved, needed, and important. Tell them of your concern about the stroke survivor. Reassure them that it isn't anyone's fault. Let them know who will take care of them when you are at the hospital. Remain available for their questions and answer them in simple terms. Explain the consequences of the stroke. Reassure them that their bodies will not have a stroke just because their loved one did. Keep hospital visits short. Remind the children of the stroke and its effects before each visit to the survivor. Ask for support and help from friends of the family who may be willing to care for children for a time. Children need to have a positive place in the family unit. In their simple, direct, no-nonsense and nondiscriminatory way, they can be wonderful teachers!

YOUR LOVED ONE has had a stroke. Hopefully, with time he or she will make the transition to rehabilitation and eventually return home. All rehabilitation hospitals or units stress motivation and independence as a priority, and cooperation is a valuable component. Help by building confidence and self-esteem in your loved one. Take advantage of stroke support groups— at these meetings you will find people with whom you can share ideas and find acceptance and encouragement. They will help you remember that stroke recovery takes time, commitment, and resourcefulness.

Therapists will have helped the stroke patient relearn or adapt to start working toward his or her potential. The caregiver or partner will probably have received an information packet from the rehabilitation hospital during the loved one's stay. This information provides insight on terms that are used, as well as a guide to outside sources such as the American Heart Association, the American Stroke Association, and the National Stroke Association. There are also many computer Web sites that may assist you and your loved one: the Stroke Network, the National Aphasia Association, and the Stroke Foundation, just to name a few. The hospital social worker can help the family expedite paperwork or channel them in the right direction. The survivor may qualify for community, county, state, or federal programs depending on the extent of his or her disability.

Aphasia, if your loved one has it, may be around for a long time. I still occasionally run into it when I tire or try to do too much at once. It can also affect the processing of thoughts, making them incomplete. Long lectures or directions are too intense for the aphasic. It is also difficult to listen, write, and process at the same time. Instead, keep it short, draw a map, or tape record a meeting. A person experiencing aphasia may mix letters, words, or adages around. Later, when the aphasia starts to recede, don't ask the person to read aloud, recite, or sing lyrics in public until he or she is ready or volunteers.

Many things that are taken for granted become a cherished gift. Reading may be affected because of comprehension difficulties. Books on Tape may be an alternative. A stroke survivor may want to write a letter or note but can't remember the process of how to do it. He or she may not remember names, common objects, or items of clothing. The survivor's speech inflection and pitch level may be affected. Along with an inability to decipher sounds, the person who has had a stroke may not be aware of his or her vocal problem. This is caused by damage to the language area of the brain.

Many stroke survivors are unaware of injuries to their affected side, or the temperature of water and other stimuli on that side of the body. For example, a few years after my stroke, while climbing to reach my pet cockatiel that had flown to the top of the valance, I fell. Feeling no pain on the left side of my body, affected by the stroke, I went about my day. That afternoon, I walked three blocks to the store. On the way home, with heavy groceries in hand, I realized my left foot wasn't supporting me. So there I was, trying to get home, carrying groceries, and hopping on one foot. I went to the emergency room to get it checked out. The X rays revealed that when I fell, I fractured my left foot. Again, in the spring of 1997, I fell from a bicycle and broke my left ankle in three places and spent most of the summer in a nursing home.

The good news is that broken bones can heal. The bad news is that a brain cannot be repaired so easily. When brain cells die it is permanent. The best hope is that new brain pathways will form to make up for those damaged, and this will happen only by trying new skills.

Concentration may be very short term. Survivors may cry at inappropriate times. They may dress only one side of their bodies, forgetting the affected side. All these conditions will become incredibly frustrating for survivors. There may also be some personality changes. Remind them of just one thing they can do now that they couldn't do when they first had the stroke. Show them you care about them. Sometimes that reassurance may come in just a touch, a hug, a smile, or silent time together.

Grief is also a phase of the recovery process. The survivor and caregiver have lost something, and both have every right to grieve. If the grief becomes depression it is a problem that must be addressed with the family and psychologists that are familiar with stroke.

Caregivers must learn to give care to themselves too. Caregivers are not expected to sacrifice their entire life for that of the survivor. The survivor does not want pity or sacrifice. The survivor needs the caregiver to survive too!

A schedule of daily activities is vitally important for survivors. It may be difficult for survivors to remember whether they have eaten, what they have eaten, or even to feel hungry. Therefore, it is a good idea to have a set mealtime. Allow the survivor to be part of the meal preparation as much as possible. A calendar and a bold-faced clock will also be helpful in keeping track of dates, time, and schedules.

To help with personal care, a chair placed in the bathtub or shower is very helpful. Learn how to assist the survivor so he or she feels confident.

If the survivor is not able to communicate, talk to him or her anyway. Tell the person what you are about to do. Be involved in therapy sessions to learn how to transfer the survivor safely.

Financial matters may be too difficult for the person who has had a stroke. For a while, the stroke survivor may not be able to count coins or understand the difference in currency, and this can cause problems. Therapy may help with this difficulty, but the caregiver can also help at home when the survivor is ready.

Let's talk about intimacy also. Far too often this important subject goes unmentioned. We are human, whole, and sexual beings. Don't be afraid to openly discuss your concerns with your physician. The survivor may feel isolated, alone, and unlovable. Survivors need the comfort of spouses or significant others now more than ever. As they recover, their sexual desires and libido may be very much alive. Their mates may feel apprehensive, or fear that by having sexual intercourse they will provoke another stroke. This is not the case. Remember, love begins in the heart. When both partners are ready and relaxed, they can enjoy intimacy again. Positions may have to change, along with improved communication regarding feeling and touch. A small night-light may be necessary so the stroke survivor can see his or her affected side in order to smoothly relocate it. Many stroke survivors and their partners can, and do, go on to live active sexual lives again.

The bad news is that stroke is the third leading cause of death in the United States and the number one cause of serious long-term disability. The good news is that there are 3.8 million stroke survivors, and that the survival rate for stroke has increased 23.5 percent in the last ten years.

My son had a telephone conversation shortly after my stroke, and if I remember it correctly, it went something like this: "Hello, I've got some bad news. Mom has had a stroke. But the good news is she's alive!" That insight from a 14-year-old is astounding. I am very much alive. Thanks to God, physicians, therapists, nursing staff, and medical technology, I'm alive to be a mother, a grandmother, a sister, an aunt, a consumer, and a homeowner. For me, there is a very full life after stroke.

LOUIS R. CAPLAN, M.D.

Positive Aspects of Rehabilitation and Illness

CLEO'S STROKE HAS not been totally negative. A wise adage says, "When life gives you lemons, make lemonade." Try to extract some gain even from setbacks. Cleo tells us how the strokes have helped her adapt and use various new strategies. The illness has made her much more self-conscious and aware of daily activities and their risks. The stroke has also made her aware of how important these daily activities are: before, they were just taken for granted; now they are much more appreciated.

The strokes have clearly taught Cleo about many activities and functions that she knew and thought little about before. They have made her much more health conscious. She is working hard to prevent further illness and vascular disease and now knows and cares more about these things than before. Health, a state of being that used to be taken for granted, is now a cherished blessing to be preserved as much as possible. The prospect of having life and abilities snatched away has also made Cleo appreciate nature's blessings more. She now tries to enjoy each day's offerings with more appreciation and gives thanks more often than she had before. Why does it take a disaster to make us live life the way we should? In many ways, Cleo has emerged from the strokes and complications a stronger, more giving, more self-conscious, and more caring person. Even severe illnesses can have some positive effects.

Advice to Patients, Families, and Future Patients (All of Us)

1. *Become knowledgeable about your body and how it works.* Health information is important if you are to become a partner in taking the best care of yourself.

2. *Know your own and family members' risk factors.* Genetics and the environment play a large role in health and disease. Learn the history of illnesses and conditions in your family. Learn which risks are likely to be genetically determined, meaning that you have a higher likelihood than average of developing those conditions. High blood pressure, diabetes, high cholesterol, and breast cancer are just a few examples of conditions that have a tendency to be more common than normal among certain families and close relatives.

3. *Learn the best way to reduce your risks and those of the family.* Many risk factors are modifiable by good health practices, such as regular exercise, avoidance of tobacco, and moderation in eating and drinking. For other conditions, such as diabetes, hypertension, and high cholesterol, monitoring by a physician may allow early detection and treatment.

4. *Learn the warning signs of important diseases such as coronary artery disease, stroke, cancer, and diabetes.* Knowledge is a powerful weapon. Early detection is most important in fighting disease effectively. Let your doctor know about any worrisome symptoms. Observe symptoms carefully so that you can explain them to your doctors.

5. *Be a strong partner in your own health care, advocating for yourself and your family for the best medical care.* Medical care has become very complex in this managed-care era. Learn as much as you can from doctors and reliable literature about your disease and risks. Don't be afraid to question your health care providers about their care. Choose providers and plans carefully. Are your doctors and nurses good listeners? Do they give you time? Are they genuinely concerned about you and your health? Are they willing to discuss their evaluation, diagnoses, and treatment with you? Are they receptive to your input? Does your doctor take a thorough history and perform a thorough examination? Does he or she know you and your family as individuals?

6. Don't hesitate to insist on consultation and care by specialists relating to your condition. Medical care has become very specialized. New technologies appear every day; new treatments are studied and reported daily. It is impossible for a primary care doctor, even an exceptionally competent one, to become up-to-date in all diseases. Seek specialty opinions especially if your condition warrants one, or is not common.

Comments About Stroke

1. *CT scans have important limitations.* Patients and many emergency doctors, primary care physicians, and internists rely much too heavily on CT scans in diagnosing and treating stroke patients. CT scans are often normal early after symptoms begin, and may remain negative even in patients with strokes. More important, CT scans (without CT angiography) do not show the blood vessels that supply the brain. Stroke is a vascular disease, and diagnosis and treatment depend on recognizing the abnormalities in the blood vessels that caused the TIAs and strokes.

2. *If you have symptoms that indicate that you might be having a stroke, seek care in a stroke center.* As must be evident from my remarks in the text, stroke is a complicated condition. Knowledge, technology, and research regarding stroke have moved so quickly during the last decade that even experts in the field have difficulty keeping up. The brain and stroke are so complex, and the technology used to evaluate stroke patients is so specialized, that stroke care requires specialists who are constantly working in the field. I believe that patients and their families deserve no less when it comes to their brains. The brain is the center of every person's humanity, and any loss of its capacity steals a little from our sense of self; no other part of the body is of such profound importance. Strokes

should be taken care of by specialists, usually neurologists, when these consultants are available.

Accurate diagnosis and treatment require that medical centers have expert stroke care readily available twenty-four hours a day, seven days a week. This necessitates experienced stroke doctors, modern technology, and systems in place to ensure that stroke patients are handled quickly and effectively. Not many hospitals live up to these requirements, but most large cities have stroke centers in one or more hospitals. Learn the name and location of the center in your city or region. Now some community hospitals, although they may not have all the personnel and equipment on site, may be connected to stroke centers by telecommunication.

Centers specializing in stroke have the necessary technology available and individuals experienced in performing and interpreting the tests. The tests can and should be done quickly. A colleague of mine said that time = brain. He meant that the longer a region is deprived of normal blood flow and nutrients, the more likely will be the death of that region. The faster brain ischemia is corrected, the more chance there is of saving vital tissues. Patients must come to centers that are equipped to diagnose and treat them quickly, and doctors must pursue the investigations and treatment urgently. Insist that you or your loved one be taken to a stroke center and not just the nearest hospital if the medical condition permits.

Public Education

Cleo wants to teach readers about stroke. Of course, that is exactly why I agreed to coauthor this book—to try to teach readers more about stroke and the brain. Understanding an illness can be very therapeutic. Public awareness about stroke can go a long way toward stroke prevention. Louis Pasteur commented that it is far better to prevent an illness from happening than to try to treat it after it occurs.

Public awareness about stroke is disproportionately weak compared with awareness about heart disease and cancer. One of the reasons is complexity. Cancer and heart disease are much simpler and more homogeneous than stroke. The brain is a much more diverse and complex organ than the heart. Also, less money, time, and effort have been expended to teach the public about stroke.

Biology and health education are critical and should be started early. Our children must know more about the body and its functions than older generations did. This goes for science in general. Science and mathematics education in the United States is generally very weak compared with that in Europe and some parts of Asia. We need more money and support for public education. More education should be provided in the workplace, in the media, and in doctors' offices and hospitals.

Advocacy for Specialty Care

Cleo identifies and describes a wide array of specialists who participated in her care—cardiologists, heart surgeons, neurologists, infectious-disease specialists, and probably others not mentioned specifically. Specialized medicine made the United States the uncontested world leader in medical research, medical education, and medical care during the last half of the twentieth century. Much of the gain related to proliferation of physicians in all fields who had great expertise and experience with certain complex problems and complex illnesses. Somehow in our health care system, individuals who had those illnesses seemed to get to the correct specialists. As those specialists saw and cared for more and more patients who had the illnesses in which they specialized, more knowledge was gained. These specialists disseminated this new knowledge in journals available to other specialists and general physicians, helping to spread the word, and treatment of patients with those illnesses improved as a result. Young doctors from all over the world came to the United States to learn from these specialists and to bring the information back to their own countries and to their own patients. In our managed care system, specialist care is threatened. Today, insurance companies dominate the scene. These companies are interested in money, the bottom line, and dividends—not primarily the health and welfare of their clients. They may think that they save money by denying care and denying consultation and treatment with specialists. The bank accounts and stock holdings of insurance and managed care executives have grown greatly while our health care becomes more and more mediocre. The government also plans to cut back funds for specialist training and specialized care. The public must be aware of this disastrous trend and fight it. All of us are potential patients. We cannot allow the destruction of one of the most effective health care systems in the world by capricious politicians and insurance company executives. Excellent health care is important and expensive, and all citizens are entitled to it. We must find a way to pay for it.

Advocacy for Nurses

Nurses are now called on to have more administrative roles, and unfortunately, administrative duties include frequent and detailed nursing notes. These administrative and recording activities take nurses away from the bedside, where they can do the most good. Nurses are also expensive professionals who more than earn their compensation. The push by insurance companies to decrease expenses, as well as the economic pressures on hospitals, has caused many hospitals to react by drastically cutting their nursing personnel. Hospitals become less humane and less safe places for ill patients when there are too few bedside nurses assigned for care. We need to advocate for nurses. They are absolutely indispensable.

Stroke Treatment Now and Tomorrow

DOCTORS' EFFORTS HAVE most recently focused on treatment of stroke patients within the first few hours after symptoms are noticed. By giving acute treatment, they strive to prevent brain infarction and so limit brain damage and disability. Newer ways of opening blocked arteries physically and chemically are being explored and studied in treatment trials. Another focus of research is administering to stroke patients various substances ("neuroprotectants") that could make nerve cells less vulnerable to reduced blood supply. It is hoped that this might allow more time for doctors to intervene to increase circulation to the threatened regions of the brain. Efforts will continue to try to get stroke patients to stroke centers that have the most experienced stroke doctors and the most advanced technology as soon as possible after stroke symptoms begin.

Even more important is a new emphasis on *recovery* after stroke. Despite the best efforts, it will never be possible to completely prevent brain damage from stroke, because no method will deliver all stroke patients to stroke centers in time, and the treatments given will not be 100 percent effective in preventing brain damage. Doctors are excited about the development of a new technique, called functional MRI (fMRI), that can show when and where the brain changes its activity after a stroke or after individuals are asked to perform a task such as moving a hand, reading words, or listening to music. When someone performs such a task, the brain areas used become more active. Increased activity requires more fuel—so more blood. An analogy is the need to step on the gas pedal to make a car go faster. The increase in blood flow increases the signal on MRI scans. When the scan taken before the task is subtracted from the scan taken during the task, the brain area that was active during the performance of the task is shown. The drawings on the facing page depict fMRI scans from

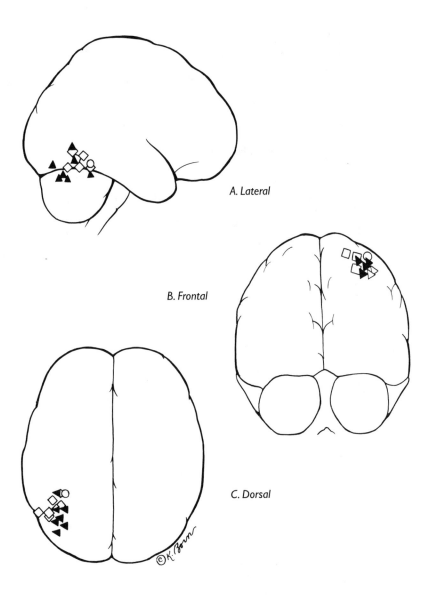

A. Lateral

B. Frontal

C. Dorsal

©K.Born

Language areas. *These drawings depict frames from a functional MRI experiment in which a group of individuals looked at a word. The areas of the brain that became active while they viewed the word are shown as dark triangles. Drawing A is a picture from the side, while B is taken from in front of the head, and C is taken from above the head. The area that is activated is in the back of the left side of the brain near the visual area.*

patients asked to look at words. A region in the left cerebral hemisphere near the visual cortex and Wernicke's area is activated.

By studying many stroke patients during recovery, doctors can learn how much and when new brain regions are activated during various body activities. Then different treatment strategies, such as medications, natural substances, magnetic stimulation, various speech therapies, acupuncture, forcing activity of weak limbs, can be studied and compared to discover whether and to what extent they improve recovery or whether they slow or impair recovery. With fMRI, new strategies can be tried and studied objectively. This new field offers great promise for stroke patients of the future.

Appendices

Suggested Reading

BrainWork: The Neuroscience Newsletter
The Dana Foundation
745 Fifth Avenue, Suite 900
New York, NY 10151
http://www.dana.org

Journal of Stroke and Cerebrovascular Diseases (professional publication)
http://www.strokejournal.org

Stroke (a professional journal of The American Heart Association)
http://intl-stroke.ahajournals.org

Stroke Smart
The National Stroke Association
9707 East Easter Lane
Englewood, CO 80112
(800) 787-6537

The Stroke Connection
The American Stroke Association (a division of the American Heart Association)
7272 Greenville Avenue
Dallas, TX 75231
(800) 553-6321

BOOKS

After a Stroke: A Support Book for Patients, Caregivers, Families, and Friends
By Geoffrey Donnan, M.D., and Carol Burton
Berkeley, Calif.: North Atlantic Books, 1990

Caplan's Stroke: A Clinical Approach, 3rd edition (professional reference)
By Louis R. Caplan, M.D.
Woburn, Mass.: Butterworth-Heineman Publications, 2000

How to Prevent a Stroke
By Peggy Jo Donahue and the Editors of *Prevention* Magazine
        (Mark L. Dyken, M.D., and Philip A. Wolf, M.D. Medical Advisers)
Emmaus, Penn.: Rodale Press, 1989

I Want to Thank My Brain for Remembering Me
By Jimmy Breslin
Boston: Little Brown & Co., 1996

I Had a Stroke and Survived
By Jerry W. Haggard
Salt Lake City: Northwest Publishing, 1994

My Year Off: Recovering Life After a Stroke
By Robert McCrum
New York–London: W. W. Norton, 1998

One-Handed in a Two-Handed World
By Tommye-Karen Mayer
Boston: Prince Gallison Press, 1996

Out of the Blue: One Woman's Story of Stroke, Love, and Survival
By Bonnie Sherr Klein, in collaboration with Persimmon Blackbridge
Berkeley, Calif.: Wildcat Canyon Press, 1997, 1998

Stroke: The Complete Guide to Recovery and Rehabilitation
By Laura Swaffield
Northamptonshire, England: Thorsons Publishing Group, 1999

Stroke: The Condition and the Patient
By John E. Sarino, M.D., and Martha Taylor Sarino
New York: McGraw-Hill, 1969

The AHA Family Guide to Stroke Treatment, Recovery, and Prevention
By Louis R. Caplan, M.D., Mark L. Dyken, M.D., and J. Donald Easton, M.D.
New York: Times Books, 1994

The Dana Guide to Brain Health
Floyd E. Bloom, M.D., M. Flint Beal, M.D., and David Kupfer, M.D., editors
New York: The Free Press, 2003

The Effective Clinical Neurologist, 2nd edition (professional reference)
By Louis R. Caplan, M.D.
Woburn, Mass.: Butterworth-Heinemann Publications, 1991

The Invaluable Guide to Life After Stroke: An Owner's Manual
By Arthur Josephs
Long Beach, Calif.: Amadeus Press, 1992

The Stroke Recovery Book: A Guide for Patients and Families
By Kip Burkman
Omaha: Addicus Books,Inc., 1996

When Someone You Love Has a Stroke
By Marilyn Larkin and Lynn Sonberg
New York: Dell Publishing Co., 1995

VIDEOS

The Brain: Our Universe Within
Discovery Channel
(Three VHS tapes)
Bethesda, Md.: Discovery Communications, Inc., 1994

The Secret Life of the Brain
PBS series; David Grubin, Producer
(Also, the companion book of the same name by Richard Restak, M.D.;
The Dana Press and Joseph Henry Press, 2001)

Resources for Patients and Families

ABLEDATA

ABLEDATA is a federally funded project whose mission is to provide information on assistive technology and rehabilitation equipment.

8630 Fenton Street, Suite 930

Silver Spring, MD 20910

Phone: 800-227-0216 (voice); 301-608-8912 (TTY)

Fax: 301-608-8958

E-mail: *abledata@macroint.com*

http://www.abledata.com

Agency for Health Care
Policy & Research Publication Clearinghouse

P.O. Box 8547

Silver Spring, MD 20907-8547

(800) 358-9295

Brain Aneurysm/AVM Center

The Brain Aneurysm/AVM Center section of the Massachusetts General Hospital/Harvard Medical School neurosurgical site includes extensive information about assessment and treatment of neurovascular problems of the brain and spinal cord, including stroke, aneurysm, and arteriovenous malformation. The site includes links to stroke sites helpful to patients, families and caregivers, and health professionals.

http://neurosurgery.mgh.harvard.edu/neurovascular/

Clearinghouse on Disability Information

330 C Street SW, Room 3132

Washington, D.C. 20202-2524

(202) 205-8241

Courage Center

Offers recreational opportunities, driving courses, stroke groups, educational material and seminars.

3915 Golden Valley Road

Golden Valley, MN 55422-9984

(612) 520-0520

http://www.courage.org/

George Washington University HEATH Resource Center

2121 K Street, NW, Suite 220

Washington, DC 20037

(800) 544-3284

(202) 973-0904

Fax: (202) 973-0908

http://www.heath.gwu.edu

Life Services for the Handicapped

352 Park Avenue South, Suite 703

New York, NY 10010-1709

(212) 532-6740

Mayo Division of Cerebrovascular Disease

http://www.mayo.edu/cerebro/education/stroke

Minnesota Senior Federation

12 offices throughout Minnesota. Offers Senior Partners Care medical program for those 65 and older, or disabled, who qualify in Minnesota.

555 Park Avenue Suite 110

St. Paul, MN 55103

(800) 365-8765

Mobility International USA

P.O. Box 10767

Eugene, OR 97440

(541) 343-1284

National Aphasia Association

400 East Thirty-Fourth Street, Room RR306

New York, NY 10016

(800) 922-4622

http://www.aphasia.org

National Council on Aging

(Family Caregivers Program)

600 Maryland Avenue SW, West Wing 100

Washington, D.C. 20024

National Institute of Neurological Diseases and Stroke

"Stroke: Hope Through Research" is a comprehensive booklet, available on-line from the National Institute of Neurological Disorders and Stroke. It provides scientific but accessible material on stroke prevention, stroke warning signs and risk factors, and emergency treatment for stroke, as well as on research, treatment, and recovery.

Office of Communications and Publications

P.O. Box 5801

Bethesda, MD 20824

(800) 352-9424

http://www.ninds.nih.gov/health_and_medical/pubs/stroke_hope_through_research.htm

National Rehabilitation Information Center for Independence

400 Forbes Boulevard, Suite 202

Lanham, MD 20706

(800) 346-2742

http://www.naric.com ("naric" is the National Association of Rehabilitation Centers, which maintains the site.)

Peterson Press

Offers presentations on stroke awareness and recovery issues from a survivor's perspective, a monthly newsletter, and other publications on healing.*http://www.PetersonPress.com*

TheRamblings@aol.com

The American Stroke Association

(A Division of the American Heart Association)

7272 Greenville Avenue

Dallas, TX 75231

(800) 553-6321

http://www.strokeassociation.org, or www.stroke.org

E-mail: *strokeconnection@heart.org*

The Brain Attack Coalition

This is a group of professional, voluntary and governmental entities dedicated to reducing the occurrence, disabilities and death associated with stroke. The goal of the coalition is to strengthen and promote the relationships among its member organizations in order to help stroke patients or those who are at risk for a stroke.

The Brain Attack Coalition Web site (*http://www.stroke-site.org/*) is maintained by the NINDS staff. The structure and ongoing content management of the site are directed and evaluated by the coalition Web site subcommittee. The following organizations are members of the coalition:

American Academy of Neurology

American Association of Neurological Surgeons

American Association of Neuroscience Nurses

American College of Emergency Physicians

American Society of Neuroradiology

American Stroke Association, a Division of the American Heart Association

Centers for Disease Control and Prevention

National Association of EMS Physicians

National Institute of Neurological Disorders and Stroke

National Stroke Association

Stroke Belt Consortium

Veterans Administration

The Brain Matters Stroke Initiative

A professional and public education program developed by the American Academy of Neurology (AAN). Although this site is really directed at stroke prevention and treatment, it has some information that might be useful for stroke victims and their families and friends.

http://www.strokematters.com

The Department of Vocational/Rehabilitative Services

Assists people with disabilities with job retraining, education, employment readiness and job placement. The local listing can be found in the telephone book government section of most cities.

http://www.ssa.gov/work/Advocates/advocates.html

The National Library Service for the Blind and Physically Handicapped (NLS)

A government agency that provides free books on tape and cassette player to those who qualify.

1291 Taylor Street, NW

Washington, D.C. 20542

The National Stroke Association

The National Stroke Association focuses on stroke prevention, treatment, rehabilitation, research, and support for stroke survivors and their families. Their site includes a stroke facts and risk factors quiz, a guide to lay and professional materials available from the association, and a regional list of stroke centers and support groups.

9707 East Easter Lane

Englewood, CO 80112-3747

(800) 787-6537

http://www.stroke.org

E-mail: *info@stroke.org*

The Stroke Center

The Stroke Center provides the latest information about stroke research and care gathered from published accounts, meeting presentations, internet searches, and direct correspondence. This easy-to-use site also provides a list of stroke facilities located in St. Louis, Missouri, and a national stroke trials directory.

http://www.strokecenter.org

The Stroke Network

A Web-based stroke support and referral organization.

http://www.strokenetwork.org

Visiting Nurse Association of America

11 Beacon Street, Suite 910

Boston, MA 02108

(617) 523-4042

INTERNATIONAL STROKE
ASSOCIATIONS AND FOUNDATIONS

Canadian Stroke Consortium

National Headquarters
1131A Leslie Street, Suite 205
Toronto, Ontario M3C 3L8
Canada
(416) 386-0844
1-866-386-0844
(416) 386-0855
E-mail: *cscheadquarters@strokeconsortium.ca*
http://www.strokeconsortium.ca

National Stroke Research Institute (NSRI)

Level 1, Neurosciences Building
Gate 10, Banksia Street
Heidelberg West
Victoria, Australia 3081
Melways: Ref 31 G4

Stroke Clubs International

805 12 Street
Galveston, TX 77550
(409) 762-1022
E-mail: *strokeclubs@aol.com*
USA

Stroke Survivors International

Internet format for stroke support.
http://www.strokesurvivors.org

The Heart and Stroke Foundation of Canada

http://www.heartandstroke.ca

The National Stroke Foundation
Suite 304, Level 3, 167 Queen Street
Melbourne 3000
Australia
http://www.strokefoundation.com

The Stroke Association in the United Kingdom
Registered Head Office
Stroke House, 240 City Road
London EC1V 2PR
England
Telephone: 020 7566 0300

The Stroke Foundation of New Zealand
National Office
P.O. Box 12482
L1, Federation House
95–99 Molesworth Street
Wellington
New Zealand
E-mail: *strokenz@stroke.org.nz*

COMPUTER SOFTWARE

Bungalow Software
Recover Speech & Language after stroke or brain injury
(800) 891-9937
Fax: (503) 526-0305
Bungalow Software, Inc.
E-mail: *info@BungalowSoftware.com*
http://www.BungalowSoftware.com

Index